HPLC and CE

Principles and Practice

HPLC and CE
Principles and Practice

Andrea Weston
Dionex
Sunnyvale, California

Phyllis R. Brown
Department of Chemistry
University of Rhode Island
Kingston, Rhode Island

Academic Press

San Diego   London   Boston   New York   Sydney   Tokyo   Toronto

Front cover illustration by James Treubig, Jr.

This book is printed on acid-free paper.

Copyright © 1997 by W. B. Saunders Company

All Rights Reserved.
No part of this publication may be reproduced or transmitted in any form or by any means, electronic or mechanical, including photocopy, recording, or any information storage and retrieval system, without permission in writing from the publisher.

Academic Press
a division of Harcourt Brace & Company
525 B Street, Suite 1900, San Diego, California 92101-4495, USA
http://www.apnet.com

Academic Press Limited
24-28 Oval Road, London NW1 7DX, UK
http://www.hbuk.co.uk/ap/

Library of Congress Cataloging-in-Publication Data

HPLC and CE : fundamentals and applications / [edited] by Andrea Weston, Phyllis R. Brown.
    p.   cm.
Includes index.
ISBN 0-12-136640-5 (alk. paper)
1. High performance liquid chromatography. 2. Capillary electrophoresis. I. Weston, Andrea, date. II. Brown, Phyllis R.
QD79.C454H63   1997
543'.0871--dc21   96-49198
                                                  CIP

PRINTED IN THE UNITED STATES OF AMERICA
97 98 99 00 01 02 EB 9 8 7 6 5 4 3 2 1

This book is dedicated to the memory of Dr. J. Calvin Giddings for his outstanding work in analytical chemistry and his unique contributions to the field of separations science. By devising the theoretical framework for chromatography, he made possible the development of high-performance liquid chromatography, which in turn opened doors to a new era of science, especially the birth of biotechnology.

Contents

Preface xiii

CHAPTER 1 High-Performance Liquid Chromatography
1.1 Introduction 1
1.2 Classification of Liquid Chromatographic Methods 2
    1.2.1 Classification According to the Mechanism of Retention 2
    1.2.2 Classification According to Operating Method 3
1.3 Basic Concepts of HPLC 7
    1.3.1 Capacity Factor 8
    1.3.2 Selectivity 9
    1.3.3 Resolution 11
    1.3.4 Band Broadening and Column Efficiency 12
1.4 Summary of Major Concepts 20
References 21

CHAPTER 2 Separations in High-Performance Liquid Chromatography
2.1 Introduction 24
2.2 Normal-Phase Chromatography 24
    2.2.1 Mechanism of Retention 25
    2.2.2 Stationary Phases for Normal-Phase Chromatography 25
    2.2.3 Mobile Phases for Normal-Phase Chromatography 26
    2.2.4 Practical Considerations 27
    2.2.5 Applications for Normal-Phase Chromatography 28
2.3 Reversed-Phase Chromatography 29
    2.3.1 Mechanism of Retention 29
    2.3.2 Stationary Phases for Reversed-Phase Chromatography 30
    2.3.3 Mobile Phases for Reversed-Phase Chromatography 31

	2.3.4 Practical Considerations	32
	2.3.5 Applications for Reversed-Phase Chromatography	37
2.4	Hydrophobic-Interaction Chromatography	38
2.5	Ion-Exchange Chromatography	38
	2.5.1 Mechanism of Retention	40
	2.5.2 Stationary Phases for Ion-Exchange Chromatography	41
	2.5.3 Mobile Phases for Ion-Exchange Chromatography	43
	2.5.4 Practical Considerations	44
	2.5.5 Applications for Ion-Exchange Chromatography	44
2.6	Size-Exclusion Chromatography	46
	2.6.1 Mechanism of Retention	46
	2.6.2 Stationary Phases for Size-Exclusion Chromatography	47
	2.6.3 Mobile Phases for Size-Exclusion Chromatography	48
	2.6.4 Practical Considerations	49
	2.6.5 Applications for Size-Exclusion Chromatography	50
2.7	Affinity Chromatography	50
	2.7.1 Mechanism of Retention	51
	2.7.2 Stationary Phases for Affinity Chromatography	52
	2.7.3 Mobile Phases for Affinity Chromatography	55
	2.7.4 Practical Considerations	55
	2.7.5 Applications for Affinity Chromatography	56
2.8	Chiral Separations	57
	2.8.1 Mechanism of Retention	57
	2.8.2 Stationary Phases for Chiral Chromatography	58
	2.8.3 Mobile Phases for Chiral Chromatography	61
2.9	Summary of Major Concepts	62
	References	67

CHAPTER 3 Instrumentation for High-Performance Liquid Chromatography

3.1	Introduction	71
3.2	Solvent Delivery Systems	71
	3.2.1 Pumps	72
	3.2.2 Minimization of Pump Pulsations	76
	3.2.3 Optimization of Flow Reproducibility	77
3.3	Solvent Degassing	78
	3.3.1 External Vacuum Degassing	78
	3.3.2 Helium Sparging	78
	3.3.3 On-line Degassing	79

3.4	Isocratic versus Gradient Pumping Systems	80
	3.4.1 Isocratic Pumping Systems	80
	3.4.2 Gradient Pumping Systems	80
3.5	Sample Introduction	83
3.6	Columns	85
	3.6.1 Stationary Phases	85
	3.6.2 Pore Size	87
	3.6.3 Particle Size	88
	3.6.4 Internal Diameter of the Column	88
	3.6.5 Column Length	88
	3.6.6 Construction Materials of the Column	89
	3.6.7 Column Care and Use	89
3.7	Detectors	90
	3.7.1 Detector Properties	90
	3.7.2 Absorbance Detectors	91
	3.7.3 Fluorescence Detectors	98
	3.7.4 Electrochemical Detectors	102
	3.7.5 Mass Spectrometric Detectors	105
	3.7.6 Other Detectors	108
3.8	Sample Preparation	109
	3.8.1 Off-line Sample Preparation Techniques	109
	3.8.2 On-line Sample Preparation Techniques	115
3.9	Troubleshooting	117
3.10	Summary of Major Concepts	129
References		130

CHAPTER 4 Capillary Electrophoresis

4.1	Introduction	134
4.2	Classification of Electrophoretic Modes	134
	4.2.1 Classification According to the Electrolyte System	135
	4.2.2 Classification According to the Contribution of the Electroosmotic Flow	136
4.3	Basic Concepts of Capillary Electrophoresis	136
	4.3.1 Electrophoretic Mobility	137
	4.3.2 Velocity	138
	4.3.3 Electroosmotic Flow	138
	4.3.4 Separation Efficiency	143
	4.3.5 Resolution	150
4.4	Summary of Major Concepts	150
References		152

CHAPTER 5 Separation in Capillary Electrophoresis

5.1 Introduction	154
5.2 Capillary Zone Electrophoresis	155
5.2.1 Mechanism of Separation	155
5.2.2 Buffer Systems	155
5.2.3 Practical Considerations	157
5.3 Micellar Electrokinetic Capillary Chromatography	161
5.3.1 Mechanism of Separation	161
5.3.2 Buffers	163
5.3.3 Practical Considerations	164
5.4 Capillary Gel Electrophoresis	165
5.4.1 Mechanism of Separation	166
5.4.2 Sieving Media	168
5.4.3 Practical Considerations	169
5.5 Capillary Electrochromatography	170
5.5.1 Mechanism of Separation	170
5.5.2 Buffer Systems	171
5.5.3 Practical Considerations	171
5.6 Chiral Capillary Electrophoresis	172
5.6.1 Mechansim of Separation	172
5.6.2 Buffer Systems	172
5.6.3 Practical Considerations	174
5.7 Capillary Isoelectric Focusing	174
5.7.1 Mechanism of Separation	174
5.7.2 Buffer Systems	177
5.7.3 Practical Considerations	177
5.8 Capillary Isotachophoresis	178
5.8.1 Mechanism of Separation	179
5.8.2 Buffer Systems	179
5.8.3 Practical Considerations	181
5.9 Summary of Major Concepts	182
References	183

CHAPTER 6 Instrumentation of Capillary Electrophoresis

6.1 Introduction	185
6.2 High-Voltage Power Supply	186
6.3 Sample Injection	187
6.3.1 Hydrostatic Injection	187
6.3.2 Electromigration Injection	187
6.3.3 Sample Concentration	189

6.4 Capillaries 190
    6.4.1 Untreated Fused Silica Capillaries 191
    6.4.2 Functionalized Fused Silica Capillaries 192
    6.4.3 Extended Path-Length Capillaries 192
6.5 Detection 193
    6.5.1 Detector Properties 194
    6.5.2 Absorbance Detectors 194
    6.5.3 Fluorescence Detectors 196
    6.5.4 Electrochemical Detectors 199
    6.5.5 Mass Spectrometry Detectors 202
6.6 Fraction Collection 205
6.7 Sample Preparation 205
6.8 Troubleshooting 206
6.9 Summary of Major Concepts 210
References 211

CHAPTER 7 Data Manipulation

7.1 Introduction 214
7.2 Identification of Peaks 214
    7.2.1 Use of Retention Data 214
    7.2.2 Standard Addition or Spiking 216
    7.2.3 Internal Standard 216
    7.2.4 Isotopic Labeling 217
    7.2.5 Enzyme Peak Shift 217
    7.2.6 Use of Ultraviolet and Mass Spectometry Libraries 220
7.3 Quantitation 222
    7.3.1 Integration 222
    7.3.2 Calculation 232
    7.3.3 Statistical Treatment of Data 236
7.4 Summary of Major Concepts 239
References 240

CHAPTER 8 Miniaturization

8.1 Introduction 242
8.2 Classification of Columns 242
8.3 Theoretical and Practical Considerations 245
    8.3.1 Theoretical Considerations 245
    8.3.2 Instrumental Considerations 248

8.4 Applications of Miniaturized Liquid Chromatography 254
8.5 Separations on a Chip 259
  8.5.1 Fabrication of Planar Devices 259
  8.5.2 Sample Introduction 262
  8.5.3 Synchronized Cyclic Capillary Electrophoresis 267
  8.5.4 Separation Efficiency 269
8.6 Summary of Major Concepts 270
References 271

Index 275

Preface

In the early 1970s, I wrote the primer *High Pressure Liquid Chromatography,* as the technique was then called. At the time, there were few commercially available instruments and a paucity of articles and books on this technique. Since then, the literature on high-performance liquid chromatography, or HPLC, has exploded and almost every laboratory that deals with analytical problems has one or more liquid chromatography instruments. Approximately 1 million liquid chromatographic analyses or separations are performed daily. The growth of HPLC has been phenomenal, and HPLC is routinely used not only in the analysis of thermally labile, nonvolatile ionic compounds but for all types of molecules from the smallest ions to macromolecules. As predicted, HPLC is indispensable in biochemistry and all the biologically related sciences. It has also been found to be a powerful analytical tool in other areas such as inorganic chemistry and organic synthesis as well as in disciplines such as environmental monitoring, clinical chemistry, oceanography, and agricultural chemistry. It is especially important as an analytical, preparative, and process technique in industries where purity of products is required by law. Most important, HPLC has opened new horizons in separations and has helped make possible the development of biotechnology where there is a great need for ultrapure products.

In the 1980s, a new separations technique, capillary electrophoresis (CE), was developed. CE created great excitement and was initially expected to replace HPLC as the method of choice for ultratrace analyses. However, it became evident that CE was complementary to HPLC and filled a different niche in separations. Since each technique has advantages and disadvantages, it is important to understand the basic theory that underlies the separations in order to choose the right technique for a problem.

Therefore, when I was asked by Academic Press to update my book, I realized that a new section on CE was mandatory. I asked a former graduate student with extensive experience in CE, Andrea Weston, to collaborate with me. By the time the book was finished, it was considerably longer than previously planned.

The book is aimed at both novices and users of HPLC and CE to give them a solid understanding of basic principles, instrumentation, methods, optimization of operation, and applications. In addition, the two techniques are compared so that users can rationally choose the appropriate technique or techniques for their analytical problems. As in my first book, we have included many illustrations to make the material easily understandable. Although this book is aimed at HPLC and CE practitioners, it can also be used for upper-level undergraduate or graduate courses. Moreover, it will be valuable for individual students, especially those from other scientific disciplines, who must use separations but have never studied the fundamentals of chromatography or electrophoresis.

Andrea and I thank all our colleagues who have helped us along the way, especially our husbands for their support, encouragement, and patience during the preparation of this book.

Phyllis R. Brown
Kingston, Rhode Island

CHAPTER 1

High-Performance Liquid Chromatography

1.1 Introduction

"Chromatography is a method in which the components of a mixture are separated on an adsorbent column in a flowing system."[1] The adsorbent material, or stationary phase, first described by Tswett[1] in 1906, has taken many forms over the years, including paper,[2] thin layers of solids attached to glass plates,[3,4] immobilized liquids,[5] gels,[6] and solid particles packed in columns.[7] The flowing component of the system, or mobile phase, is either a liquid or a gas. Concurrent with development of the different adsorbent materials has been the development of methods more specific to particular classes of analytes.[6,8–10] In general, however, the trend in development of chromatography has been toward faster, more efficient systems.[11–20]

Liquid chromatography (LC), which is one of the forms of chromatography, is an analytical technique that is used to separate a mixture *in solution* into its individual components. As indicated by Tswett, the separation relies on the use of two different "phases" or "immiscible layers," one of which is held stationary while the other moves over it. Liquid chromatography is the generic name used to describe any chromatographic procedure in which the mobile phase is a liquid. The separation occurs because, under an optimum set of conditions, each component in a mixture will interact with the two phases differently relative to the other components in the mixture. High-performance liquid chromatography (HPLC) is the term used to describe liquid chromatography in which the liquid mobile phase is mechanically pumped through a column that contains the stationary phase. An HPLC instrument, therefore, consists of an injector, a pump, a column, and a detector.

This chapter introduces the basic theory and terminology governing chromatographic separations and the equations used to calculate the effectiveness of the analytical system. With this information, the best separation mechanism and column characteristics for a given problem can be chosen,

1

on the basis of the nature of the components in the mixture as well as the physical and chemical characteristics of the column.

1.2 Classification of Liquid Chromatographic Methods

There are two ways to classify liquid chromatographic methods. The first and more common classification is based on the mechanism of retention, and from this the chromatographic modes discussed in Chapter 2 are derived. For example, the normal-phase mode can be performed by taking advantage of either the adsorption mechanism or the partition mechanism. The gel-filtration mode is performed using the mechanism of size exclusion. The second classification discussed below is based on the separation principle and is found mostly in the literature published before the 1990s.

1.2.1 Classification According to Mechanism of Retention

The most popular classification scheme stems from the manner in which the analyte interacts with the stationary phase. With this approach, chromatography may be divided into five separation mechanisms: adsorption, partition, size exclusion, affinity, and ion exchange, as illustrated in Figure 1.1.

Adsorption chromatography is based on competition for neutral analytes between the liquid mobile phase and a neutral, solid stationary phase. The analytes interact with the stationary phase according to the premise "like likes like": polar solutes will be retained longest by polar stationary phases, and nonpolar solutes will be retained best by nonpolar stationary phases. In adsorption chromatography the solute molecules are in contact with both the stationary phase and the mobile phase, simultaneously. Under these conditions, the solutes are said to be in an anisotropic environment.

Partition chromatography is also based on competition for neutral analytes, but in this case the stationary phase is considered to be a neutral liquid. Owing to the instability of liquid stationary phases, true partition

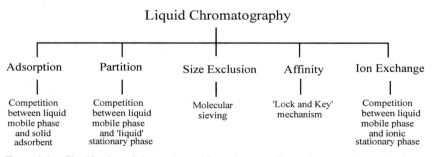

Figure 1.1 Classification of chromatographic modes according to the retention mechanism.

chromatography is not commonly used in modern HPLC. Instead, long-chain (C_{18}) "bonded-phase" columns have been developed in which the long alkyl chains are considered to behave like a liquid. Thus, the process is termed partition when the solute is transferred from the bulk of one phase into the bulk of the other, so that the solute molecules are completely surrounded by molecules of one phase. Under these conditions, the solutes are said to be in an isotropic environment.

Ion-exchange chromatography (IEC) is based on the principle that opposites attract. Ion-exchange chromatography is used to separate charged analytes and therefore occurs as a result of interaction between a charged solute and an oppositely charged, solid stationary phase. Ion-exchange chromatography can be applied to any solute that can acquire a charge in solution. Thus, even carbohydrates, which are largely uncharged below pH 12, can be separated by ion-exchange chromatography at sufficiently high pH.

Size-exclusion chromatography (SEC) is based on the sieving principle. In SEC, the stationary phase particles are manufactured with a wide range of pore sizes, causing the stationary phase to behave like a molecular sieve. As a result of the sieving action, the solutes are separated on the basis of size, with the larger ones eluting first (BOCOF, big ones come out first).

Affinity chromatography is based on the lock-and-key mechanism prevalent in biological systems. The retention mechanism is very specific, but the technique is more time-consuming and more expensive than those employing other retention mechanisms.

The mechanisms described above form the basis for the chromatographic modes described in Chapter 2, namely, normal-phase, reversed-phase, size-exclusion, ion-exchange, and affinity chromatographies. However, other modes that are variations of those mentioned above, such as hydrophobic-interaction chromatography (HIC), chiral, ion-exclusion, and ion-pair chromatographies are also used and will be mentioned.

1.2.2 Classification According to Operating Method

The second classification scheme is less common than the first but is found in the literature. It is based on the operating method, or the mechanism by which the sample is removed from the column, and is therefore dependent on the nature of the mobile phase. This classification, which was introduced by Tiselius[21] in 1940, includes elution development, displacement development, and frontal analysis, as shown in Figure 1.2.[22] In practice, only elution and to a lesser extent displacement development are commonly used.

In elution development a small volume of sample is introduced onto the head of the column, and the components are adsorbed onto the stationary phase to various degrees. The solutes are eluted from the column using

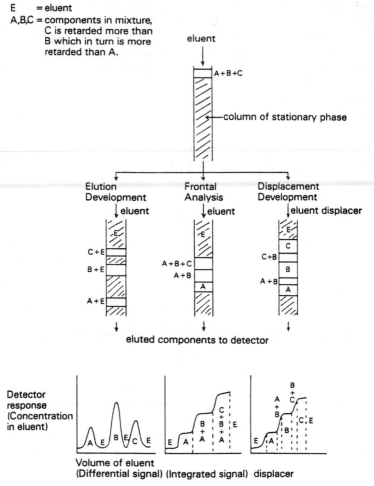

Figure 1.2 Classification of chromatographic methods according to the operating method. (Reprinted from Ref. 22 with permission.)

a mobile phase which has a greater affinity for the sample components than for the stationary phase. Since the components can be completely separated with a zone of mobile phase between them, elution chromatography is commonly used for analytical separations where quantitation and characterization may be important.

Elution chromatography may be subclassified according to the "continuity" of the mobile phase. Isocratic elution is the term used when the sample is introduced onto the column and eluted from it under the same set of mobile phase conditions. Isocratic elution is the most common way to remove solutes from a column. It is suitable when the components have similar affinities for the stationary phase and are therefore eluted rapidly, one after the other.

Gradient elution is most commonly performed with reversed-phase chromatography but is also used with other modes such as ion-exchange chromatography. It involves a continuous change in the composition of the mobile phase to achieve separation of sample components of widely varying affinities for the stationary phase. A weakly eluting mobile phase is used at the start of the run, and eluting strength is increased over the course of the separation. This allows for sufficient resolution of the early, weakly retained solutes while ensuring that the elution time of the later peaks is not excessively long. A binary gradient refers to a gradient employing two different eluents, a ternary gradient refers to a system where three different eluents are used, and a quaternary gradient is one where four different eluents are used. The solvent composition gradient can be linear, convex, or concave, as illustrated in Figure 1.3, and may involve a change in concentration, pH, polarity, or ionic strength.

To develop a gradient separation, the appropriate strengths for the different eluents must be determined. Often the best approach is to start with a linear binary gradient, beginning with 100% of the weak eluent (e.g., water) and ending with 100% (e.g., acetonitrile, ACN) of the strong eluent. However, the two solutions must be completely miscible at the concentrations used.

Figure 1.4 shows the development of a separation of a mixture of alkylphenones using different binary gradients.[23] Figure 1.4a shows a linear

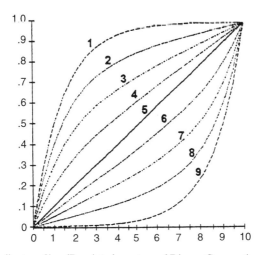

Figure 1.3 Gradient profiles. (Reprinted courtesy of Dionex Corporation, from the DX500 pump manual.)

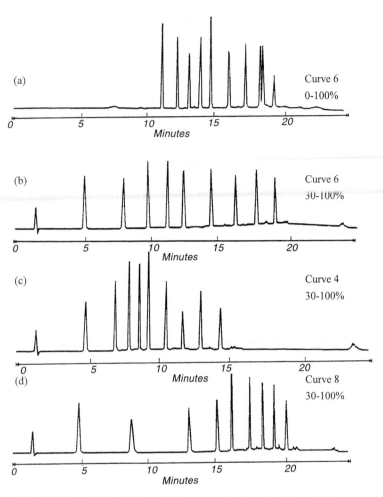

Figure 1.4 Gradient elution of nine alkylphenones showing the effect of different gradient profiles on the peak resolution: (a) linear, 0–100% acetonitrile (ACN), (b) linear, 30–100% ACN, (c) convex, 30–100% ACN, and (d) concave, 30–100% ACN. Curve numbers refer to profiles of Figure 1.3. (Adapted from Ref. 23 with permission.)

gradient starting at 100% water and ending with 100% ACN in 12 min. After 12 min, the gradient was held at 100% ACN for a further 5 min to allow time for all the peaks to be eluted. At 17 min the system was returned to initial conditions to re-equilibrate the column with the weak eluent prior to the start of the next run. Although all peaks are eluted under these conditions, no peaks appear in the first 10 minutes, indicating that the initial conditions are too weak.

Figure 1.4b shows a linear gradient separation of the same sample using the same two eluents except that the starting conditions have been changed from 0% ACN to 30% ACN. In this case, the separation is good, but it might be possible to decrease the run time by changing the gradient profile. Figure 1.4c shows the effect on resolution and run time of using a convex gradient profile, whereas a concave profile is shown in Figure 1.4d. The convex profile is the most appropriate because not only is the resolution still good, but the run time has been decreased by 5 min. If an adequate separation were not obtained, the strong solvent can be changed. If this still were to produce an inadequate separation, a ternary gradient or even a quaternary gradient may be appropriate, but the greater complexity makes gradient optimization much more difficult.

Step elution is similar to gradient elution in that the composition of the mobile phase changes during the separation process. However, in step elution the change is not continuous. It involves a sudden change in the composition of mobile phase followed by a period where the mobile phase is held constant. This procedure results in a mobile phase profile that resembles a series of steps.

Displacement chromatography is commonly used for preparative-scale separations, but, because of its "focusing" or concentrating effect, it also shows potential on the analytical scale, for example, for the concentration of minor components in complex mixtures.[24,25] Operationally, displacement chromatography is similar to the step elution process, except that in the displacement process the mobile phase has a greater affinity for the stationary phase than for the sample components, and therefore the components are eluted ahead of the displacer front. The focusing effect of displacement chromatography is due to the fact that the concentration of the displacer determines the concentration of the product bands.[26]

Frontal analysis is a preparative method, used primarily for the separation of one readily eluted component from the other, more tightly held components. The technique is performed by the continuous addition of a sample mixture onto the column. Initially, the component of interest, that is, the component with the least affinity for the stationary phase, will pass through the column while the other sample components are retained to various degrees by the stationary phase. As a result of the continuous sample application, the concentration of bound components steadily builds up at the head of the column. When the column capacity for any given component is exceeded, that component also passes through the column. Therefore, the first component is eluted from the column initially as a pure band and subsequently as a mixture with the next components to be eluted.

1.3 Basic Concepts of HPLC

Chromatography is described and measured in terms of four major concepts: capacity, efficiency, selectivity, and resolution. The capacity and selectivity of the column are variables that are controlled largely by the column manufacturer, whereas efficiency and resolution can be controlled,

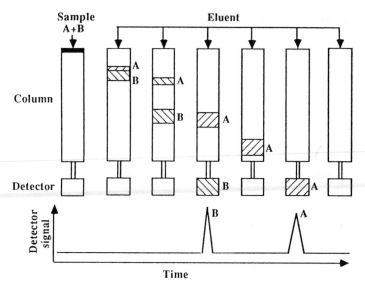

Figure 1.5 Separation of two components, A and B. (Reprinted from Ref. 27 with permission.)

to some extent, by the chromatographer. To obtain the best possible separation, the efficiency of the chromatographic system must be optimized in order to minimize band broadening (Fig. 1.5).[27] The column should have the capacity to retain the solutes, and it should have the appropriate selectivity to resolve the analytes of interest.

1.3.1 Capacity Factor

For effective liquid chromatographic separations, a column must have the capacity to retain samples and the ability to separate sample components, efficiently. The capacity factor, k'_R, of a column is a direct measure of the strength of the interaction of the sample with the packing material and is defined by the expression

$$k'_R = \frac{t_R - t_0}{t_0} = \frac{V_R - V_0}{V_0} \tag{1.1}$$

where t_R is the time taken for a specific solute to reach the detector (retention time) and t_0 is the time taken for nonretained species to reach the detector (holdup time). These terms are illustrated in Figure 1.6. The same value for k' is obtained if volumes are used instead of times: V_R is the volume of solution that is pumped through the detector before a specific peak is eluted (retention volume), and V_0 is the volume of solvent pumped through the detector between the time of injection and the appearance of the nonretained species (void volume). The void volume is equal to the volume of the column not occupied by packing material.

The capacity factor of a column is mostly a function of the packing material but can be manipulated to a degree by varying the solvent strength. The higher the capacity factor of the column, the greater is its ability to retain solutes. Using a column with a higher capacity factor is often the best way to improve the resolution of a separation. Because a higher capacity factor will also result in longer analysis times, a compromise between resolution and analysis time must be reached. Typically, a k' value between 2 and 5 represents a good balance between analysis time and resolution; however, k' values between 1 and 10 are usually acceptable.

1.3.2 Selectivity

The selectivity of the chromatographic system is a measure of the difference in retention times (or volumes) between two given peaks and describes how effectively a chromatographic system can separate two compounds (Figure 1.7). Selectivity is usually defined in terms of α, where

$$\alpha = \frac{t_2 - t_0}{t_1 - t_0} = \frac{V_2 - V_0}{V_1 - V_0} = \frac{k'_2}{k'_1} \qquad (1.2)$$

The selectivity of a column is primarily a function of the packing material, although the chromatographer has some control using the mobile phase or temperature. The value for α can range from unity (1), when the retention times of the two components are identical ($t_2 = t_1$), to infinity if the first component of interest is eluted in the void volume. If α is approaching 1,

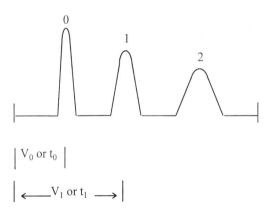

Figure 1.6 Chromatographic terms used to calculate column capacity.

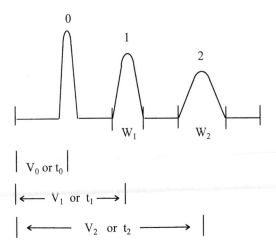

Figure 1.7 Chromatographic terms used to calculate selectivity.

then regardless of the number of theoretical plates or the length of time the components stay on the column, there will be no separation.

The most powerful approach to increasing α is to change the composition of the mobile phase. If changing the concentration of the components in the mobile phase provides insufficient change, altering the nature of one of the components will often be sufficient. Figure 1.8[28] shows the effect on the separation of acetonaphthalene and dinitronaphthalene of changing the mobile phase from 23% dichloromethane/77% pentane to 5% pyridine/

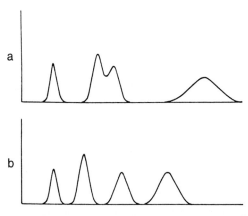

Figure 1.8 Effect on selectivity of changing the composition of the mobile phase: (a) 23% dichloromethane/77% pentane and (b) 5% pyridine/95% pentane. (Reprinted from Ref. 28 with permission.)

1.3 Basic Concepts of HPLC

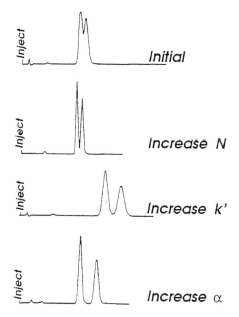

Figure 1.9 Effect of selectivity, capacity factor, and efficiency on resolution.

95% pentane. The α value changes from 1.05 when using dichloromethane to 2.04 when using pyridine.

1.3.3 Resolution

Resolution is a term used to describe the degree of separation between neighboring solute bands or peaks. It is affected by the selectivity (α), efficiency (N) and capacity (k') of the column. The resolution equation [Eq. (1.3)] describes the relationship between those factors and indicates how they can be manipulated in order to improve the resolution between two peaks.

$$R = \frac{1}{4} \frac{\alpha - 1}{\alpha} (N^{1/2}) \frac{k'}{1 + k'} \qquad (1.3)$$

The effect of selectivity, capacity factor, and efficiency on resolution is illustrated in Figure 1.9. Typically, an R value greater than 0.8 is required for accurate quantification of two peaks. A value of 1, for two equally sized peaks, indicates an overlap of about 2% for one band over the other.[29] Chromatograms of two peaks of unequal sizes for given resolution values are shown in Figure 1.10.[30]

The most effective way to alter resolution is to change the selectivity

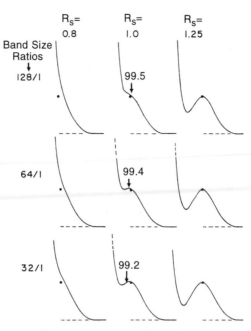

Figure 1.10 Standard resolution curves for band-size ratios of 32/1, 64/1, and 128/1 and R_s values of 0.8–1.25. (Reprinted from Ref. 30 with permission.)

or the capacity factor of the column. The effect of increasing the efficiency of the column by increasing the column length or flow-rate velocity is less significant, as resolution increases proportionally as the square root of the number theoretical plates. Thus, doubling the number of theoretical plates by adding a second column increases resolution by only a factor of 1.4. If increased resolution is required, a column with a higher capacity factor is often the best choice. However, increasing the capacity factor will increase the analysis time, so a compromise must be reached between resolution and analysis time.

1.3.4 Band Broadening and Column Efficiency

When a sample mixture is first applied to the head of a column, as illustrated in Figure 1.5, the width of the sample is very narrow. By the time the components are eluted from the end of the column, however, the band widths have broadened. This phenomenon occurs because, as the sample mixture moves down the column, the various sample components interact with, and are retained to various degrees by, the stationary phase. This interaction, along with the torturous path of the sample components through the packing material, causes the increase in band width, a process known as band broadening. The amount of band broadening determines, to an extent, the degree to which two components can be separated; thus, band broadening should be kept to a minimum.

The efficiency of a column is a number that describes peak broadening as a function of retention, and it is described in terms of the number of theoretical plates, N. Two major theories have been developed to describe column efficiency, both of which are used in modern chromatography. The plate theory, proposed by Martin and Synge,[31] provides a simple and convenient way to measure column performance and efficiency, whereas the rate theory developed by van Deemter et al.[32] provides a means to measure the contributions to band broadening and thereby optimize the efficiency.

(i) Plate Theory

The empirical expressions derived in the plate theory are generally applicable to all types of column chromatography. Although the relationships are valid only for Gaussian peak shapes, for convenience they are also generally applied to nonsymmetrical peaks. The major assumption in the plate theory is that there is an instantaneous equilibrium set up for the solute between the stationary and mobile phases. The main criticisms of the plate theory are that it does not consider the effects of band broadening on separation, nor does it consider the influence of chromatographic variables such as particle size, stationary phase loading, eluent viscosity, and flow rate on column performance.[22]

In the chromatographic model proposed in the plate theory, the chromatographic column is considered to consist of a number of thin sections or "plates," each of which allows a solute to equilibrate between the stationary and mobile phases. The greater the number of theoretical plates (N), the more efficient the column is considered to be. The movement of a solute along the column is viewed as a stepwise transfer from one theoretical plate to the next. The thinner the theoretical plates, the greater the number that can be envisaged within a given length of column. These terms are related as follows:

$$H = L/N \qquad (1.4)$$

where L is the length of the column (millimeters). Thus, the smaller the height equivalent to a theoretical plate (HETP, or H), the greater is the efficiency of the column. In general, the H value is smaller for small stationary phase particle sizes, low mobile phase flow rates, less viscous mobile phases, higher separation temperatures, and smaller solute molecule sizes.

Efficiency, N, is defined in terms of the retention time (t_R) of the solute, measured at the peak apex, and the standard deviation, σ, of the solute population in the peak measured as the peak width:

$$N = (t_R/\sigma)^2 = t_R^2/\sigma^2 \qquad (1.5)$$

where σ for a Gaussian peak is given by

$$\sigma = \frac{w_{50}}{2.345} = \frac{w_T}{4} = \frac{w_{4.4}}{5} \qquad (1.6)$$

as shown in Figure 1.11, and w is the peak width at different heights on the curve. From the 5σ table, N can be calculated in a number of ways, depending on where the width is measured. The most commonly used method for the calculation of N is the tangent method, owing to its relative simplicity, but the 5σ method provides the greatest sensitivity to peak tailing. Because of the relative insensitivity to peak tailing, the inflection, peak at half-height, and 3σ methods should not be used.

$$\text{Peak half-height:} \quad N = 5.54\left(\frac{t_R}{w_{50}}\right)^2 \qquad (1.7)$$

$$\text{Tangent method:} \quad N = 16\left(\frac{t_R}{w_T}\right)^2 \qquad (1.8)$$

$$5\sigma \text{ method:} \quad N = 25\left(\frac{t_R}{w_{4.4}}\right)^2 \qquad (1.9)$$

The efficiency can be varied by changing physical column parameters such as the length, diameter, and construction material of the container of the column. It can also be varied by changing chemical parameters such

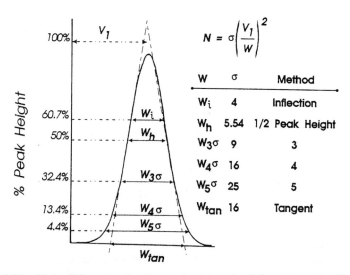

Figure 1.11 Methods for estimating the standard deviation (σ), and 5σ table. (Reprinted from Ref. 23 with permission.)

as the size of the particles constituting the packing material or the mobile-phase velocity.

(ii) Rate Theory and Band Broadening

There are three predominant mechanisms for transport of a solute through a chromatographic column: (1) convective transport in the mobile phase as it flows between the particles in the column, (2) diffusive transport through stagnant pools of liquid in the column packing, and (3) perfusive transport through the porous particles. The rate theory developed by van Deemter et al.,[32] and later modified by others,[33-35] considers the diffusional factors that contribute to band broadening in the column (column variance, σ^2_{col}) and avoids the assumption of an instantaneous equilibrium inherent in the plate theory.

In its most general form, the van Deemter equation may be written

$$H = A + B/\mu + C\mu \quad (1.10)$$

where H represents the efficiency of the column and μ represents the average linear velocity of the mobile phase. The A term represents the contribution to band broadening by eddy diffusion, the B term represents the contribution from longitudinal diffusion, and the C term represents the contribution from resistance to mass transfer. Diffusion is not restricted to transport of solute particles through stagnant pools of liquid in the stationary phase, however, but also occurs as the solute is carried by convective transport between the particles in the column. Huber[36] found that there were at least four terms that should be considered to describe column efficiency adequately, and the contribution of these factors to efficiency is described in the modified van Deemter equation:

$$H = A + B/\mu + C_s\mu + C_m\mu. \quad (1.11)$$

In Eq. (1.11) the C terms represents the contributions to zone broadening from resistance to mass transfer in the stationary phase and the mobile phase, respectively.

Because H represents the column variance or band broadening, the value for H should be kept to a minimum. One way to determine the experimental conditions that will give minimum zone dispersion and maximize efficiency is provided by the use of a van Deemter plot. A van Deemter plot, as shown in Figure 1.12, is a graph of plate height versus the average linear velocity of the mobile phase. The data are determined experimentally using measured values for retention time, void volume or dead time, and peak width to determine N and hence H at various flow rates. According to the plot, at flow rates below the optimum the overall efficiency is dependent on diffusion effects (the B term). At higher flow rates the efficiency decreases because the mass transfer, or C, terms become more important. In the plot shown in Figure 1.12, the A term is a constant,

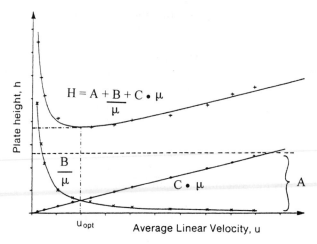

Figure 1.12 Hypothetical van Deemter plot showing the relationship between efficiency and average linear velocity of the mobile phase.

independent of flow rate. Despite the reduced efficiency at higher flow rates, it is common to operate the system at high flow rates to save time and operating costs; thus, the plot may also be used to determine the best conditions to minimize the analysis time consistent with an acceptable value for H.

A. Eddy Diffusion As a solute molecule passes through the column, it can follow a variety of different paths around the stationary-phase particles, as illustrated in Figure 1.13. Each of the paths will be of a different length, so that as solute molecules of the same species follow different paths, they will arrive at the outlet of the column at different times. This form of diffusion is known as eddy diffusion and is represented by the A term of the van Deemter equation.

In practice, the solute molecules are not fixed in a single path but can diffuse laterally into other channels, thus decreasing the contribution to

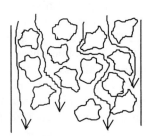

Figure 1.13. Eddy diffusion.

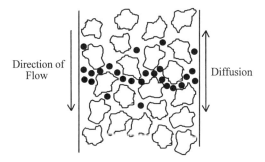

Figure 1.14   Longitudinal diffusion.

band broadening from eddy diffusion. This band broadening process is dependent completely on the stationary phase and is independent of the flow rate of the mobile phase. It can be minimized if the column is packed uniformly with particles of constant size.

B. Longitudinal Diffusion In chromatography, the sample mixture ideally travels through the column as tight zones of individual sample components separated by zones of mobile phase, or as regions of high solute concentration separated by regions of high solvent concentration. Whenever a concentration gradient exists, however, diffusion of molecules will occur, from a region of high concentration to a region of low concentration (Fig 1.14). This form of diffusion is known as longitudinal diffusion and is represented by the B term of the van Deemter equation. It is related only to the mobile phase and is independent of the stationary phase.

Longitudinal diffusion occurs in all directions; the molecules at the front of the zone will move forward into the next zone, the molecules at the end of the zone will fall back into the previous zone, and diffusion will also occur toward the column walls. As diffusion is a time-dependent process, the longitudinal diffusion effect increases at low mobile-phase flow rates.

C. Resistance to Mass Transfer In the plate theory, it was assumed that the transfer of solute molecules between the mobile phase and the stationary phase was instantaneous. In the rate theory, it is accepted that there is a finite rate of mass transfer. In addition, molecules of the same species may spend different lengths of time in the stationary and mobile phases (Fig. 1.15). Resistance to mass transfer is represented by the C term of the van Deemter equation.

If the time required for mass transfer is much greater than the time required for the solute molecules to flow over the surface of the packing material, some molecules in the convective stream can move down and out of the column before diffusion of others into and out of the centers of

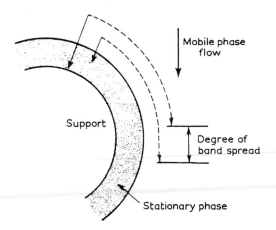

Figure 1.15 Stationary-phase mass transfer.

the particles can be completed. Thus, with conventional porous packing materials, increasing the flow rate reduces resolution and capacity.

D. Stagnant Mobile Phase Stagnant mobile-phase mass transfer has been identified as one of the major contributors to peak dispersion in liquid chromatography[28,37,38] and is also represented by the C term of the van Deemter equation. As shown in Figure 1.16, the presence of immobile

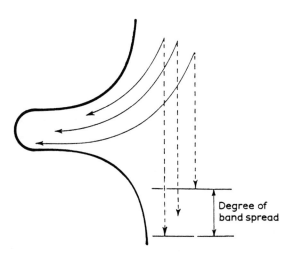

Figure 1.16. Stagnant mobile-phase mass transfer.

solvent, trapped either between particles of packing material or in the pores within the particles provides the means for solute particles to diffuse right into the stagnant pool in the pore, effectively becoming trapped. Thus, column variance may be reduced by using short columns, uniformly packed with small-diameter particles of constant size.

E. Extracolumn Band Broadening or Variance To maximize the effective number of theoretical plates, the contribution of the entire chromatographic system to band broadening (system variance, σ_{sys}^2) must be minimized. The system variance may be broken down into contributions from the column variance, σ_{col}^2, as described above, and extracolumn diffusion and mixing processes, σ_{ex}^2. As with the case of the column variance, extracolumn variance is an additive property and may be broken down into the major components:

$$\sigma_{ex}^2 = \sigma_{inj}^2 + \sigma_{det}^2 + \sigma_t^2 \quad (1.12)$$

where σ_{inj}^2 is the injection volume variance, σ_{det}^2 is the detector variance, and σ_t^2 is the variance contribution from the tubing in the chromatographic system. The maximum injection volume has been calculated by Guiochon and Colin[39] and is dependent on the column length and the packing material diameter.

The detector variance is a sum of the variances due to the detector cell volume and the detector time constant. The detector cell volume is dependent on the column length and the diameter of the packing material[39] and should be no greater than 0.1 times the peak band volume (4σ).[40] As the detector cell volume becomes larger, the efficiency of the column, N, decreases. This effect is most dramatic for the early eluting peaks. Because the solute passes through the detector cell quickly, the detector time constant must be fast enough to allow the detector to respond to the brief presence of the solute in the flow cell. If the time constant is too high, resolution and sensitivity are compromised,[41] as illustrated in Figure 1.17. The setting of the time constant is dependent on the length of the column and the diameter of the packing material.[42]

The final contribution to extracolumn variance, namely, the contribution from the tubing in the chromatographic system, is a function of both the length and more importantly the radius of the tubing.[43] At high flow rates and with short lengths of narrow-bore tubing, less band broadening is seen than would have been expected;[44,45] thus, in practice, when a well-constructed chromatographic system is used with a column containing uniformly packed, small-diameter particles, the most significant contribution to band broadening comes from the variance arising from mass transfer effects.

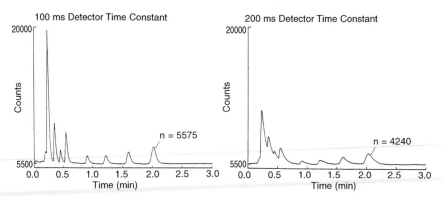

Figure 1.17 Effect of detector time constant on resolution, system efficiency, and sensitivity: (a) 100 msec, (b) 200-msec. Flow cell volume was 2.4 ml, and both chromatograms were recorded at the same sensitivity. (Reprinted from Ref. 41 with permission.)

1.4 Summary of Major Concepts

1. Liquid chromatographic methods may be classified according to either the mechanism by which analytes are retained on the column or the mechanism by which they are removed from it. The mechanism of retention classification is the most popular scheme, and five major retention mechanisms have been identified:

adsorption,
partition,
ion exchange,
affinity, and
size exclusion.

These mechanisms form the basis of the chromatographic modes discussed in Chapter 2.

2. Three methods of removing the analytes from the column are known:

elution development,
displacement, and
frontal analysis.

Isocratic elution is the most popular method for removing solutes from a column, but gradient elution is useful, especially in reversed-phase chromatography, for the removal of solutes with widely varying k' values.

3. Chromatography is described and measured in terms of four major concepts:

capacity,
efficiency,
selectivity, and
resolution.

For the best separation results, the efficiency of the system must be optimized in order to minimize band broadening, and the column should have the capacity to retain the analytes and sufficient selectivity to resolve them.

The capacity factor of a column, k', is primarily a function of the packing material but can be manipulated by varying the solvent composition and strength. The higher the capacity factor of a column, the greater is its ability to retain solutes but the longer is the analysis time; a k' value between 2 and 5 provides a reasonable compromise between analysis time and resolution.

The selectivity of a column, α, is also a function of the packing material but can be varied by changing the temperature or the nature of the mobile phase. The value for α can range from 1 if the retention times of two peaks are identical to infinity if the first peak is eluted in the void volume.

Resolution, R, is dependent not only on the selectivity and the capacity of a column but also on the efficiency of the system. Typically, an R value greater than 0.8 is required for accurate quantitation of two peaks. The most effective way to alter resolution is to change either the selectivity or the capacity of the column.

4. The efficiency of a column, N, is a number that describes peak broadening as a function of retention and is dependent on the entire chromatographic system. The most common method for calculating N is the tangent method.

Two theories have been developed to describe the efficiency of the column:

the plate theory and
the rate theory.

The plate theory assumes that an instantaneous equilibrium is set up for the solute between the stationary and mobile phases, and it does not consider the effects of diffusional effects on column performance. The rate theory avoids the assumption of an instantaneous equilibrium and addresses the diffusional factors that contribute to band broadening in the column, namely, eddy diffusion, longitudinal diffusion, and resistance to mass transfer in the stationary phase and the mobile phase. The experimental conditions required to obtain the most efficient system can be determined by constructing a van Deemter plot.

References

1. Tswett, M. S., *Ber. Dtsch. Bot. Ges.* **24**, 316 (1906).
2. Consden, R., Gordon, A. H., and Martin, A. J. P., *Biochem. J.* **38**, 224 (1944).

3. Izmailov, N. S., and Schraiber, M. S., *Farmatsiya (Sofia)* 1 (1938).
4. Stahl, E., *Pharmazie*, **11**, 633 (1956).
5. Martin, A. J. P., and Synge, R. L. M., *Biochem. J.* **35**, 1358 (1941).
6. Flodin, P., and Porath, J., *Nature (London)* **183**, 1657 (1959).
7. Majors, R. E., *Anal. Chem.* **44**, 1722 (1972).
8. Taylor, T. I., and Urey, H. C., *J. Chem. Phys.* **6**, 429 (1938).
9. Small, H., Stevens, T. S., and Bauman, W. C., *Anal. Chem.* **47**, 1801 (1975).
10. Axen, R., Ernback, S., and Porath, J., *Nature (London)* **214**, 1302 (1967).
11. Halasz, R., Endele, R., and Asshauer, J., *J. Chromatogr.* **112**, 37–60 (1975).
12. Ishii, D., Asai, K., Hibi, K., Jonokuchi, T., and Nagaya, M., *J. Chromatogr.* **144**, 157 (1977).
13. DiCesare, J. L., Dong, M. W., and Atwood, J. G., *J. Chromatogr.* **217**, 369–386 (1981).
14. Dolan, J. W., van der Wal, S., Bannister, S. J., and Snyder, L. R., *Clin. Chem.* **26**, 871–880 (1980).
15. Noda, Akio, Nishiki, and Setsuko, *Shokuhin Eiseigaku Zasshi* **18**, 321–327 (1977).
16. Bannister, S. J., van der Wal, S., Dolan, J. W., and Snyder, L. R., *Clin. Chem.* **27**, 849–855 (1981).
17. Burns, D. A., *ACS Symp. Ser.* **136**, 15–30 (1980).
18. Scott, R. P. W., Kucera, P., and Munroe, M., *J. Chromatogr.* **186**, 475–487 (1979).
19. Katz, E., and Scott, R. P. W., *J. Chromatogr.* **253**, 159–178 (1982).
20. Takeuchi, T., and Ishii, D., *J. Chromatogr.* **213**, 25 (1981).
21. Tiselius, A., *Ark. Kemi. Mineral. Geol.* **14B**, 22 (1940).
22. Braithwaite A. and Smith F. J., "Chromatographic Methods," 4th Ed. Chapman & Hall, New York, 1990.
23. "Developing HPLC Separations, Book Two." Waters Chromatography, Milford, Massachusetts, 1991.
24. Frenz, J., Bourell, J., and Hancock, W. S., *J. Chromatogr.* **512**, 299 (1990).
25. Ramsey, R., Katti, A. M., and Guiochon, G., *Anal. Chem.* **62**, 2557 (1990).
26. Frenz, J., *LC-GC* **10**, 668–674 (1992).
27. Haddad, P. R., and Jackson, P. E., *J. Chromatogr. Libr.* **46**, 2 (1990).
28. Snyder, L. R., *J. Chromatogr.* **63**, 15 (1971).
29. Snyder, L. R., and Kirkland, J. J., "Introduction to Modern Liquid Chromatography," 2nd Ed. Wiley, New York, 1979.
30. Snyder, L. R., *J. Chromatogr. Sci.* **10**, 200 (1972).
31. Martin, A. J. P., and Synge, R. L. M., *Biochem. J.* **35**, 1358 (1941).
32. van Deemter, J. J., Zuiderweg, F. J., and Klinkenberg, A., *Chem. Eng. Sci.* **5**, 271 (1956)
33. Giddings, J. C., *J. Chem. Phys.* **31**, 1462 (1959).
34. Giddings, J. C., "Dynamics of Chromatography, Part I, Principles and Theory." Marcel Dekker, New York, 1965.
35. Huber, J. F. K., and Hulsman, J. A., *Anal. Chem.* **38**, 305 (1967).
36. Huber, J. F. K., *in* "Advances in Chromatography", (A. Zlatkis, ed.), Preston. Evanston, Illinois, 1969.
37. Horvath, C., and Lin, H.-J., *J. Chromatogr.* **149**, 43 (1978).
38. Huber, J. F. K., *Ber. Bunsen-Ges. Phys. Chem.* **77**, 179 (1973).
39. Guiochon, G., and Colin, H., *J. Chromatogr. Libr.* **28**., Chapter 1 (1984).

40. Kirkland, J. J., Yau, W. W., Stocklosa, H. J., and Dilks, C. H. Jr., *J. Chromatogr. Sci.* **15**, 303–316 (1977).
41. Simpson, R. C., *in* "High Performance Liquid Chromatography" (P. R. Brown and R. A. Hartwick, eds.), *Chemical Analysis, Vol. 98.*, p. 382. Wiley, New York, 1989.
42. Kucera, P., *J. Chromatogr.* **198**, 93–109 (1980).
43. Scott, R. P. W., and Kucera, P., *J. Chromatogr.* **125**, 251–263 (1976).
44. Golay, M. J. E., and Atwood, J. G., *J. Chromatogr.* **186**, 353–370 (1979).
45. Atwood, J. G., and Golay, M. J. E., *J. Chromatogr.* **218**, 97–122 (1981).

CHAPTER 2

Separations in High-Performance Liquid Chromatography

2.1 Introduction

A variety of chromatographic modes have been developed, on the basis of the mechanisms of retention and operation introduced in Chapter 1. The key chromatographic modes are normal-phase, reversed-phase, ion-exchange, size-exclusion, and affinity chromatography. In addition to the major modes, however, there are a number of techniques that could be viewed as submodes, which are discussed briefly. For example, hydrophobic-interaction chromatography, like reversed-phase chromatography, separates species according to hydrophobicity but employs a salt gradient rather than organic solvents for elution, wheras "chiral" separations are based on a specific biological interaction in a similar fashion to affinity chromatography.

This chapter introduces the chromatographic modes and explains how they work. Examples of compounds separated by each mode, and the advantages and disadvantages, are provided to help chromatographers choose the most appropriate mode for a given group of analytes.

2.2 Normal-Phase Chromatography

The term "normal phase" is used to denote a chromatographic system in which a polar stationary phase is employed and a less polar mobile phase is used for elution of the analytes. In the normal-phase mode, neutral solutes in solution are separated on the basis of their polarity; the more polar the solute, the greater is its retention on the column. Since the mobile phase is less polar than the stationary phase, increasing the polarity of the mobile phase results in decreased solute retention.

Although normal-phase chromatography can be performed using either

2.2 Normal-Phase Chromatography

partition or adsorption mechanisms, the dominant retention mechanism is adsorption. As a consequence, normal-phase chromatography is also known in the literature as adsorption chromatography or liquid–solid chromatography. The stationary phase is polar, typically as a result of hydroxyl groups (-OH); thus, if a neutral solute molecule has a permanent dipole, or if a dipole can be induced on it, then it will be attracted by dipole–dipole interaction to the stationary phase surface. In adsorption chromatography, sample retention is directly proportional to the surface area of the stationary phase. The surface area should be kept below 400 m^2/g, however, because higher surface areas can be achieved only at the expense of smaller pores. Pores that are too small lead to poor mass transfer and lower column efficiencies.[1]

2.2.1 Mechanism of Retention

Two models have been developed to describe the adsorption process. The first model, known as the competition model, assumes that the entire surface of the stationary phase is covered by mobile phase molecules and that adsorption occurs as a result of competition for the adsorption sites between the solute molecule and the mobile-phase molecules.[1] The solvent interaction model, on the other hand, suggests that a bilayer of solvent molecules is formed around the stationary phase particles, which depends on the concentration of polar solvent in the mobile phase. In the latter model, retention results from interaction of the solute molecule with the secondary layer of adsorbed mobile phase molecules.[2] Mechanisms of solute retention are illustrated in Figure 2.1.[3]

2.2.2 Stationary Phases for Normal-Phase Chromatography

The typical stationary phases employed in normal-phase or adsorption chromatography are common porous adsorbents, such as silica and alumina,

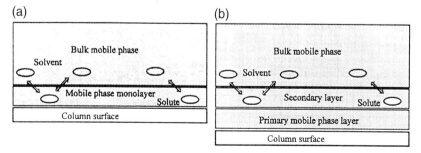

Figure 2.1 (a) Competition and (b) solvent interaction models of solute retention in normal-phase chromatography (Reprinted from Ref. 3 with permission.)

that have polar hydroxyl groups on the surface. Silica is the preferred stationary phase owing to ready availability, low cost, and known performance. For basic compounds such as amines, which are very strongly retained on silica, however, it may be advantageous to use alumina. In addition to the porous adsorbents, a variety of polar bonded phases exist in which functional groups, such as cyano [-(CH$_2$)$_3$C≡N], diol [-(CH$_2$)$_3$OCH$_2$CH(OH)CH$_2$OH], and amino groups [-(CH$_2$)$_n$NH$_2$, where n is 3 or 4],[4] are chemically bonded to the silica stationary phase. These functional groups are significantly less polar than the silanol group (-SiOH) and therefore result in less retention than is seen with the silica and alumina columns.

Despite the availability of several types of columns, one of the greatest disadvantages of normal-phase chromatography is the lack of selectivity imparted by the columns. Virtually all compounds are eluted in the same order, regardless of the column selected. Thus, changes in selectivity are achieved primarily through changing the mobile phase. In general, the order of elution is as shown in Table 2.1,[1] with highly polar compounds being retained longer.

2.2.3 Mobile Phases for Normal-Phase Chromatography

The mobile phases used in normal-phase chromatography are based on nonpolar hydrocarbons, such as hexane, heptane, or octane, to which is added a small amount of a more polar solvent, such as 2-propanol.[5] Solvent selectivity is controlled by the nature of the added solvent. Additives with large dipole moments, such as methylene chloride and 1,2-dichloroethane, interact preferentially with solutes that have large dipole moments, such as nitro- compounds, nitriles, amines, and sulfoxides. Good proton donors such as chloroform, m-cresol, and water interact preferentially with basic solutes such as amines and sulfoxides, whereas good proton acceptors such as alcohols, ethers, and amines tend to interact best with hydroxylated molecules such as acids and phenols. A variety of solvents used as mobile phases in normal-phase chromatography are listed in Table 2.2, some of which may need to be stabilized by addition of an antioxidant, such as 3–5% ethanol, because of the propensity for peroxide formation.

Table 2.1 General Order of Elution for Species Separated by Normal-Phase Chromatography

Saturated hydrocarbons < olefins < aromatic hydrocarbons ≅ organic halides < sulfides < ethers < nitro compounds < esters ≅ aldehydes ≅ ketones < alcohols ≅ amines < sulfones < amides < carboxylic acids

2.2 Normal-Phase Chromatography

Table 2.2 Mobile Phases in Normal-Phase Chromatography[a]

Solvent	Solvent strength $\varepsilon°$		Characteristic
	Silica	Alumina	
n-Hexane	0.01	0.01	Mobile phase
n-Heptane	0.01	0.01	Mobile phase
Isooctane	0.01	0.01	Mobile phase
1 Chlorobutane	0.2	0.26	Large dipole
Chloroform	0.26	0.40	Proton donor
Methylene chloride	0.32	0.42	Large dipole
Ethyl acetate	0.38	0.58	Proton donor
Tetrahydrofuran	0.44	0.57	Proton acceptor
Propylamine	~0.5	—	Proton acceptor
Acetonitrile	0.5	0.65	Dipole
Methanol	~0.7	0.95	Proton acceptor

[a] Adapted from Ref. 1 with permission.

The strength of the solvent is defined by the solvent strength parameter, $\varepsilon°$, as listed in Table 2.2. A solvent with a low $\varepsilon°$ is chosen, and quantities of a second solvent with a greater $\varepsilon°$ are added until the desired separation is achieved. If the desired separation does not result from altering the concentration of the second solvent, either the nature of the second solvent can be changed or another additive can be introduced. Readers are directed to Refs. 1, 6, and 7 for in-depth discussions on the development of mobile phases for normal-phase chromatography.

2.2.4 Practical Considerations

The major disadvantage of normal-phase separations using silica or alumina stationary phases, and to a lesser extent the bonded phases, is the effect of water. Water, absorbed from the atmosphere into the eluents, is adsorbed onto the strongest adsorption sites on the stationary phase, leaving a more uniform distribution of weaker sites to retain the sample. Water adsorption leads to a decrease in solute retention, and the adsorbent is said to become deactivated. In practice, run times are decreased the longer the eluent sits around, and they often return to normal when a new bottle of eluent is made up. It is possible to minimize the effect of absorbed water by making up an eluent 50% saturated with water. The advantages of making a saturated solution of eluent are as follows: (1) less variation in solute retention from run to run, (2) higher sample loadings, (3) higher column efficiencies and reduced peak tailing for basic compounds, and (4) reduced catalytic activity of the adsorbent.[1] However, column re-equilibration is slow when changing from one solvent to another.

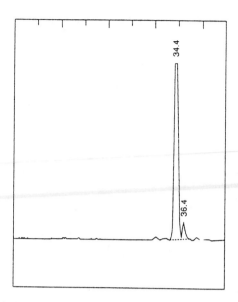

Figure 2.2 Normal-phase chromatographic analysis of abamectin 8,9-oxide. The major peak at 34.4 min is the B1a component (present at >80%), and the peak at 36.4 min is the B1b component. Chromatography conditions: column, Hypersil 3 mm silica column, 250 × 4.6 mm I.D.; mobile phase, 1.1% 3-hydroxypropionitrile in 1-chlorobutane–1,2-dichloroethane (5:2); flow rate, 1.0 ml/min; column temperature, 30°C; detection, Vis absorbance at 570 nm following postcolumn reaction; PCRS solution, 5% TCA in 1,2-dichloroethane saturated with NSA flowing at 0.5 ml/min, at 45°C. (Reprinted from Ref. 8 with permission.)

2.2.5 Applications for Normal-Phase Chromatography

Normal-phase chromatography is most commonly applied to the analysis of samples that are soluble in non-polar solvents, and it is particularly well suited to the separation of isomers and to class separations.[1] It is also possible to separate species using normal-phase chromatography on the basis of the number of electronegative atoms such as oxygen or nitrogen. Fat- and water-soluble vitamins, hydrocarbons, and pesticides have all been separated using hexane as the mobile phase. Figure 2.2 shows the use of normal-phase chromatography for the analysis of the 8,9-oxide of avermectin B1, a macrocyclic lactone with potential as an insecticide.[8]

A less obvious example of normal-phase chromatography is the separation of saccharides and oligosaccharides in foods[9,10] and in biological mixtures,[11,12] using a mobile phase consisting of acetonitrile/water or acetonitrile/dilute phosphate buffer. Although the separation mode has occasionally been misidentified as reversed phase, it is normal phase by virtue of the fact that increased aqueous levels of the mobile phase reduce carbohydrate retention, and elution order follows carbohydrate polarity.[1]

Normal-phase separations have occasionally been combined off-line with reversed-phase chromatography to separate a wider range of species than could be accomplished by either technique alone.[1] The feasibility of such a system, however, is contingent on the compatibility of the normal-phase eluent with that of the reversed-phase column.

2.3 Reversed-Phase Chromatography

Reversed-phase chromatography, the most widely used chromatographic mode,[1] is used to separate neutral molecules in solution on the basis of their hydrophobicity.[13] As the name suggests, reversed-phase chromatography (often referred to as RP chromatography) is the reverse of normal-phase chromatography in the sense that it involves the use of a non-polar stationary phase and a polar mobile phase. As a result, a decrease in the polarity of the mobile phase results in a decrease in solute retention. Modern reversed-phase chromatography typically refers to the use of chemically bonded stationary phases, where a functional group is bonded to silica, as illustrated in Figure 2.3. For this reason, reversed-phase chromatography is often referred to in the literature as bonded-phase chromatography. Occasionally, however, polymeric stationary phases such as polymethacrylate or polystyrene, or solid stationary phases such as porous graphitic carbon, are used.

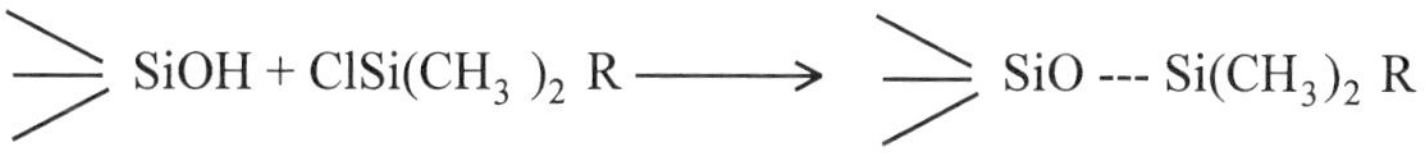

$$\equiv SiOH + ClSi(CH_3)_2 R \longrightarrow \equiv SiO - Si(CH_3)_2 R$$

Figure 2.3 Reaction of silica gel with a functional group to produce a reversed-phase stationary phase.

2.3.1 Mechanism of Retention

Two main theories, the so-called solvophobic and partitioning theories, have been developed to explain the separation mechanism on chemically bonded, non-polar phases, as illustrated in Figure 2.4. In the solvophobic theory the stationary phase is thought to behave more like a solid than a liquid, and retention is considered to be related primarily to hydrophobic interactions between the solutes and the mobile phase[14–16] (solvophobic effects). Because of the solvophobic effects, the solute binds to the surface of the stationary phase, thereby reducing the surface area of analyte exposed to the mobile phase. Adsorption increases as the surface tension of the mobile phase increases.[17] Hence, solutes are retained more as a result

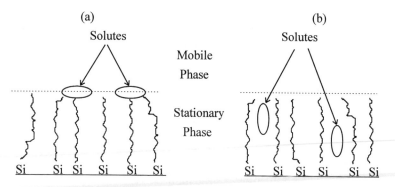

Figure 2.4 (a) Solvophobic and (b) partitioning models of solute retention.

of solvophobic interactions with the mobile phase than through specific interactions with the stationary phase.

In the partitioning model of retention, the stationary phase plays a more important role in the retention process.[18–21] The solute is thought to be fully embedded in the stationary phase chains, rather than adsorbed on the surface, and therefore is considered to be partitioned between the mobile phase and a "liquid-like" stationary phase. Although the exact mechanism of retention on chemically bonded, non-polar phases is still a matter of debate, there is general agreement that as the chain length of the bonded material increases, the retention mechanism approaches a partitioning mechanism; with shorter chain lengths, the retention mechanism becomes more similar to the adsorption mechanism.[22] Needless to say, depending on the nature of the bonded phase, the mechanism is probably a combination of both those described above.

2.3.2 Stationary Phases for Reversed-Phase Chromatography

The most common stationary phases in reversed-phase chromatography are those in which a functional group is chemically attached to a silica support (bonded phases). The most popular bonded phases are the alkyl groups, such as $-CH_3$, $-C_4H_9$, $-C_8H_{17}$, and $-C_{18}H_{37}$, phenyl ($-C_6H_5$) groups, cyano [$(-CH_2)_3CN$] groups, and amino [$(-CH_2)_3NH_2$] groups, with retention increasing exponentially with chain length. The performance of the bonded phases is determined by four factors: (1) the base silica and its pretreatment, (2) the choice of functional group, (3) the amount of material bonded to the silica (carbon load), and (4) secondary bonding reactions (end capping).

The amount of carbon introduced into the stationary phase by the functional group is referred to as the carbon load, and it is measured as a weight percentage of the bulk silica packing. The carbon load is altered by

2.3 Reversed-Phase Chromatography

changing the functional group; the higher the carbon load, the greater is the reversed-phase retention. For steric reasons, as illustrated in Figure 2.5, it is not possible for all the silanol groups on the silica surface to react with the functional groups, and usually only about 45% of the silanols will be bonded.[23] Residual, unreacted, acidic silanol groups can cause tailing of basic solutes such as amines, owing to a mixed adsorption/partition retention mechanism. Thus, unreacted silanols are often removed by treatment with a small silating agent, such as trimethylchlorosilane ($Si(CH_3)_3Cl$), a process known as end-capping.

The functional group affects not only the carbon load, but also column selectivity and efficiency; the nature of the functional group controls selectivity, while the chain length controls column efficiency. Although shorter chains result in more efficient columns, the sample capacity decreases with decreasing chain length. Thus a good general purpose RP column is one packed with 3–5 μm spherical silica and a pore size of 60–120 Å, which has C_{18} functional groups, has a carbon load of 7–10%, and is end-capped.

A few polymeric reversed-phase stationary phases are available which provide the advantage that they can be operated over a wider pH range than the silica-based columns. Polymeric columns, however, tend to be less efficient than silica-based ones and are often less retentive.

2.3.3 Mobile Phases for Reversed-Phase Chromatography

The mobile phases used in reversed-phase chromatography are based on a polar solvent, typically water, to which a less polar solvent such as acetonitrile or methanol is added. Solvent selectivity is controlled by the nature of the added solvent in the same way as was described for normal-phase chromatography; solvents with large dipole moments, such as methylene chloride and 1,2-dichloroethane, interact preferentially with solutes that have large dipole moments, such as nitro- compounds, nitriles, amines, and sulfoxides. Solvents that are good proton donors, such as chloroform, *m*-cresol, and water, interact preferentially with basic solutes such as amines

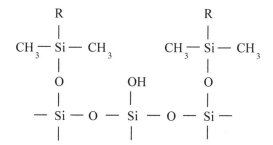

Figure 2.5 Silica surface following reaction with a functional group but prior to end capping.

and sulfoxides, and solvents that are good proton acceptors, such as alcohols, ethers, and amines, tend to interact best with hydroxylated molecules such as acids and phenols. Table 2.3 lists some useful solvents for use as mobile phases in reversed-phase chromatography.

The eluting strength of the solvent is inversely related to its polarity, as indicated in Table 2.3 where P' is the solvent polarity parameter. A solvent with a low P' is chosen initially, and quantities of a second solvent with a greater P' are added until the desired separation is achieved. More details on the development of mobile phases can be found elsewhere.[1,6,7]

2.3.4 Practical Considerations

One of the major disadvantages of bonded-phase columns is the instability of the silica support at high and low pH. Generally, mobile phases should be used within the range pH 2–8 because, below pH 2, hydrolysis of the bonded functional groups occurs, resulting in decreased retention. Above pH 8, silica dissociates and the silica support starts to dissolve. Polymeric supports, on the other hand, can be operated over the range pH 2–12, and often provide a selectivity different from that of the silica-based columns; however, the polymeric phases tend to be less efficient than silica-based columns.

The popularity of reversed-phase chromatography can be explained by its unmatched simplicity, versatility, and scope.[12] Although reversed-phase chromatography is used routinely for separating non-polar, non-ionic compounds, it is also possible and practical to separate ionic compounds on standard reversed-phase stationary-phase materials by using secondary equilibria, such as ion suppression, ion-pair formation, metal complexation, and micelle formation. To take advantage of these secondary equilibria,

Table 2.3 Mobile Phases in Reversed-Phase Chromatography[a]

Solvent	Solvent polarity P'
Water	10.2
Dimethyl sulfoxide	7.2
Ethylene glycol	6.9
Acetonitrile	5.8
Methanol	5.1
Acetone	5.1
Dioxane	4.8
Ethanol	4.3
Tetrahydrofuran	4.0
2-Propanol	3.9

[a] Adapted from Ref. 1 with permission.

additives are introduced into the mobile phase. In this way, neutral and ionic species may be separated simultaneously, and the rapid re-equilibration time of the stationary phase with changes in the mobile phase composition allows gradient elution techniques to be used routinely.

(i) Ion-Suppression Chromatography

Weak acids and weak bases, for which ionization can be suppressed, may be separated on reversed-phase columns by the technique known as ion suppression. In this technique a buffer of appropriate pH is added to the mobile phase to render the analyte neutral or only partially charged. Acidic buffers such as acetic acid are used for the separation of weak acids, and alkaline buffers are used for the separation of weak bases.[25] As a result of the added buffer, samples are eluted as sharper zones than they would have been in a similar mobile phase without the addition of a modifier.[24]

Ion suppression is not often applied to strong acids or strong bases because of the extremes in pH that would be required for retention. Suppression by the addition of buffers is restricted to the range pH 3–8, because silica-based stationary phases are unstable in solutions of either high or low pH (i.e., above pH 8 or below pH 2). These restrictions do not apply to polymeric supports, and polymer-based stationary phases can be used in the separation of a wider range of solutes using the ion-suppression technique.

(ii) Ion-Pair Chromatography

The analysis of strong acids or strong bases using reversed-phase columns is typically accomplished by the technique known as ion-pair chromatography (also commonly called paired-ion or ion-interaction chromatography). In this technique, the pH of the eluent is adjusted in order to encourage ionization of the sample; for acids pH 7.5 is used, and for bases pH 3.5 is common.[26] Retention is then altered by including in the mobile phase a bulky organic molecule[27] having a charge opposite from that of the ion to be analyzed. The counterion is the ion-pairing reagent. Three basic models have been proposed to describe the ion-pair mechanism: the ion-pair model, the dynamic ion-exchange model, and the ion-interaction model.[14,24,28]

The ion-pair model[14] postulates that because the ion-pairing reagent contains bulky organic substituents the ion pair, which is formed in the mobile phase between the solute and the ion-pairing reagent, is hydrophobic in character and will therefore adsorb onto the hydrocarbon stationary phase. The longer the alkyl chain on the pairing agent, the less polar is the ion pair, the greater is the affinity of the ion pair for the stationary phase, and the longer is its retention. The dynamic ion-exchange model[28] proposes that it is the unpaired organic counterion that adsorbs to the surface of the nonpolar stationary phase, forming a dynamic equilibrium between ion-pairing reagent in the mobile phase and ion-pairing reagent adsorbed to the surface of the stationary phase. This interaction causes the column to behave as an ion exchanger, and sample ions are therefore separated on the basis of conventional ion-exchange mechanisms.

The ion-interaction model[24] can be viewed as an intermediate between the two previous models and proposes the formation of an electrical double layer at the stationary-phase surface. As in the previous model, it is suggested that a dynamic equilibrium is established between the ion-pairing reagent adsorbed onto the stationary phase and that free in solution. However, this model proposes that to this primary layer of charge is attracted a second layer of loosely held ions of opposite charge. Transfer of solutes through the double layer to the stationary phase is then a function of both electrostatic effects and the solvophobic effects responsible for retention in reversed-phase chromatography.[25]

A list of Ion-pairing reagents is given in Table 2.4.[29] An alkyl sulfonate is a good first choice for basic solutes, whereas quaternary amines are useful for acidic solutes. Just as in reversed-phase chromatography, the most popular solvent combinations in ion-paired reversed-phase chromatography are water/methanol and water/acetonitrile. The major limitation, however, is the solubility of the ion-pairing reagent.[29] If the ion-pairing reagent is not soluble in the organic modifier, precipitation may occur within the chromatographic system. Thus, it is wise to check the solubility of the proposed ion-pairing reagent in the least polar solvent to be used.[29] Figure 2.6 shows the use of sodium octanesulfonate as the ion-pairing reagent for the determination of vitamin C in fish feed.[30] The mobile phase was a methanol/water mixture containing sodium acetate and ethylenediaminetetraacetic acid (EDTA) and adjusted to pH 4.0 with glacial acetic acid.

Retention in ion-pair chromatography is affected by the type, size, and

Table 2.4 Ion-Pairing Reagents[a]

Type	Main applications
Quaternary amines, e.g., tetramethylammonium, tetrabutylammonium, and palmityltrimethylammonium ions	For strong and weak acids, sulfonated dyes, carboxylic acids
Tertiary amines, e.g., trioctylamine	Sulfonates
Alkyl and aryl sulfonates, e.g., methane or heptane sulfonic acid, camphorsulfonic acid	For strong and weak bases, benzalkonium salts, catecholamines
Perchloric acid	Forms very strong ion pairs with wide range of basic solutes
Alkyl sulfates, e.g., lauryl sulfate	Similar to sulfonic acids; yields different selectivities

[a] Reprinted from Ref. 29 with permission.

2.3 Reversed-Phase Chromatography

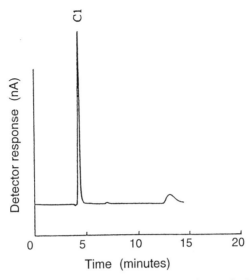

Figure 2.6 Chromatogram showing the presence of vitamin C in trout feed. Chromatographic conditions: column, Nucleosil C_{18}; flow rate, 1.0 ml/min; detection, electrochemical set at 0.75 V, 200 nA full scale. (Reprinted from Ref. 30 with permission.)

concentration of the ion-pairing reagent. The larger and more concentrated the reagent, the better able it is to form ion pairs and the longer the retention time of the solutes. Retention is also affected by the nature and concentration of the organic modifier. The more lipophilic the modifier and the more concentrated it becomes, the shorter is the retention time of the sample solutes.

(iii) Metal Complexation

In metal complexation, transition metals are added to the mobile phase to modulate selectivity. There are two approaches to the use of this technique. The first approach simply involves the introduction of a suitable metal ion, such as nickel(II), directly into the mobile phase. A solute ion that complexes rapidly with the transition metal will tend to be made more hydrophilic and will be eluted from the column more rapidly. Excellent efficiency and high selectivity have been achieved by that means.[31]

The second approach involves the addition of a chelated form of the transition metal, for example, C_{12}-dien plus zinc(II) (where C_{12}-dien is 4-dodecyldiethylenediamine), into the mobile phase (Fig. 2.7). In this example, the triamine chelate binds strongly to the metal ion, forming a complex cationic counterion. The metal chelate represents a conformationally semirigid structure with a local polarized charge center. As such, not only is there the typical electrostatic attraction of anions for the positively charged

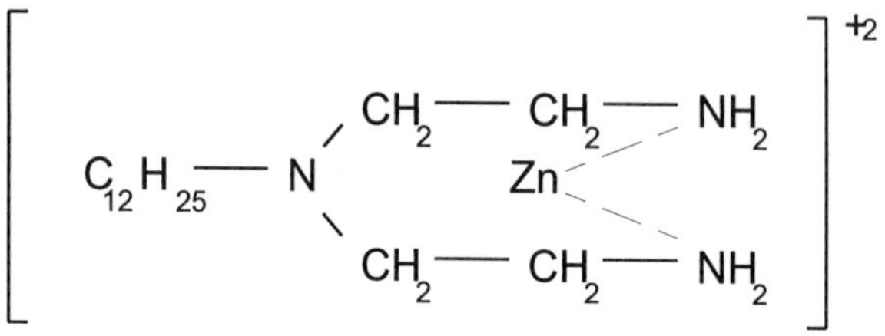

$$\left[\begin{array}{c} \quad\quad\quad CH_2 \!\!-\!\! CH_2 \!\!-\!\! NH_2 \\ C_{12}H_{25} \!\!-\!\! N \quad\quad\quad\quad Zn \\ \quad\quad\quad CH_2 \!\!-\!\! CH_2 \!\!-\!\! NH_2 \end{array} \right]^{+2}$$

Figure 2.7 Structure of C_{12}-dien–Zn(II) metal chelate complex.

chelate, but steric effects and hydrogen bonding may also occur.[32] Because the structure of the metal chelate can be tailored for favorable interaction with certain solutes of interest, this approach provides specific control of selectivity.

(iv) Micelle Formation

The "bulky organic molecules" used in ion-pair chromatography are often surfactants, present either in quantities below the critical micelle concentration (CMC) or in a mobile phase containing sufficient organic modifiers to alter or disrupt the micellar assembly.[33] Micellar chromatography is typically used for the separation of nonionic solutes, and the surfactants are used at or just above the critical micelle concentration. In addition, the mobile phase contains little or no organic modifier. Common surfactants used in micellar chromatography include dodecylsulfate, hexadecyltrimethylammonium ion, and even nonionic detergents such as Triton X-100.

Temperature has a large effect on the mass transfer between the micelle and the stationary phase and can therefore be used to improve the efficiency of the separation. Micellar chromatography should be carried out at elevated temperature, typically around 40°C. At elevated temperatures, the effects of flow rate and surfactant concentration on the efficiency of the separation are minimized. For optimum efficiency, however, the flow rate should be minimized while still maintaining a reasonable elution time.[33] Likewise, a surfactant concentration close to but above the critical micelle concentration should be used.[33]

One of the major differences between micellar chromatography and standard reversed-phase chromatography is the selectivity of the separation. As the micelle concentration is increased, solute retention decreases as a result of increased solute–micelle interactions in the mobile phase. The rate of decrease varies from solute to solute, however, since different solutes will have a different affinity for the micelles; thus, inversions in retention orders are produced.[34]

2.3.5 Applications for Reversed-Phase Chromatography

Reversed-phase chromatography is the most popular mode for the separation of low molecular weight (<3000), neutral species that are soluble in water or other polar solvents. It is widely used in the pharmaceutical industry for separation of species such as steroids, vitamins, and β-blockers. It is also used in other areas; for example, in clinical laboratories for analysis of catecholamines, in the chemical industry for analysis of polymer additives, in the environmental arena for analysis of pesticides and herbicides, and in the food and beverage industry for analysis of carbohydrates, sweeteners, and food additives.

Because the mobile phase in reversed-phase chromatography is polar, reversed-phase chromatography is suited to the separation of polar molecules that either are insoluble in organic solvents or bind too strongly to the polar, normal-phase materials.[13] As many biological molecules fall into this category, including amino acids, proteins, peptides, nucleic acids, and oligosaccharides, the development of wide-pore reversed-phase packings has dramatically affected the separation of biopolymers in the life sciences and biotechnology.[35] Figure 2.8 shows comparative tryptic maps of recombinant tissue-type plasminogen activator (rt-PA), a glycosylated protein of approximately 64,000 daltons approved by the U.S. Food and Drug Administration (FDA) for treating heart attack patients.[36] The bottom map of a mutant rt-PA is virtually identical to the reference map except that the retention time of one peptide is shifted. Such a variance in peak retention is used to identify changes in the amino acid sequence of the protein.[37]

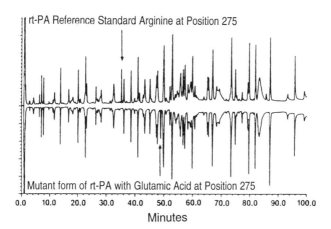

Figure 2.8 Tryptic map chromatograms of the rt-PA reference standard and a mutant form of rt-PA with a glutamic acid residue in place of the normal arginine residue at position 275. Arrows illustrate the differences in the two chromatograms caused by the substitution. (Reprinted from Ref. 36 with permission.)

2.4 Hydrophobic-Interaction Chromatography

Hydrophobic-interaction chromatography (HIC) is widely used for the separation and purification of sensitive, biologically active species, such as proteins. As with reversed-phase chromatography, solutes, on the basis of their hydrophobicity and retention, can be described in terms of the solvophobic theory.[13] In hydrophobic-interaction chromatography, however, analytes are induced to bind to a weakly hydrophobic stationary phase using a totally aqueous, buffered mobile phase of high ionic strength. Selective desorption and elution of the analyte are accomplished using a gradient of decreasing salt concentration. As a consequence of this mild interaction, there is a greater probability that the analytes will be eluted from the column with conformational structure and biological activity intact.

Retention and selectivity in HIC depend on temperature in addition to the nature of the mobile and stationary phases; the strength of the hydrophobic interaction increases with increasing temperature.[38] The primary stationary-phase variables are the ligand density, structure, and hydrophobicity. The more hydrophobic the stationary phase, the greater is the retention of the analytes. If the stationary phase is too hydrophobic, however, the analytes may become denatured. Some analytes may only be handled satisfactorily on hydrophilic stationary phases.[12]

The mobile-phase parameters that can be adjusted to optimize selectivity include the nature and concentration of the salt used in the eluent,[39] the slope of the salt gradient,[40] the eluent pH,[41] and the addition of organic solvents,[42] denaturing agents,[43] or surfactants[44] to the eluent. If single-salt gradients are used, selectivity may be varied by holding the surface tension of the eluent constant while exchanging one salt for another. In HIC, however, gradient elution with two or three salts is more efficient in modulating selectivity.[45] Figure 2.9 shows the effect of different gradients on the separation of a group of protein standards by HIC. As can be seen, the binary gradients are more effective at separating the proteins than the single-salt gradient.

The initial salt concentration is important since the higher it is, the greater is the analyte retention.[13] The effect is more pronounced for the early eluting analytes, and there is little effect on the well-retained ones. In general, higher resolution is attained with shallower salt gradients. Shallower gradients, however, also result in increased retention times, greater peak dilution, and, for some analytes, increased denaturation.

2.5 Ion-Exchange Chromatography

In ion-exchange chromatography (IEC), species are separated on the basis of differences in electric charge. The primary mechanism of retention is the electrostatic attraction of ionic solutes in solution to "fixed ions" of

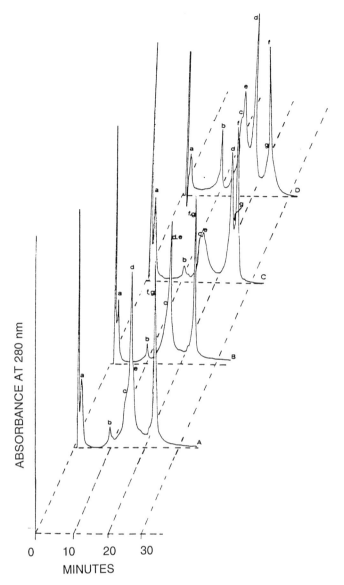

Figure 2.9 Hydrophobic-interaction chromatography of proteins. (A) Ammonium sulfate gradient from 2.16 to 0 M; (B) ammonium sulfate and tetrabutylammonium bromide gradients from 2.16 to 0 M and from 0 to 10 mM, respectively; (C) ammonium sulfate and tetrabutylammonium bromide gradients from 2.16 to 0 M and from 0 to 20 mM, respectively; (D) ammonium sulfate and tetrabutylammonium bromide gradients from 2.16 to 0 M and from 0 to 40 mM, respectively. Chromatography conditions: column, silica-bound polyether, 10 cm × 4.6 mm I.D.; temperature, 25°C; flow rate, 1 ml/min; gradient, linear for 30 min; background buffer, 50 mM phosphate, pH 6.5. Peaks: a, cytochrome c; b, ribonuclease A; c, β-lactoglobulin A; d, lysozyme; e, ovalbumin; f, α-chymotrypsinogen A; g, fetuin. (Reprinted from Ref. 45 with permission.)

opposite charge on the stationary phase support. The stationary phase or ion exchanger is classified as an anion-exchange material when the fixed ion carries a positive charge and as a cation exchanger when it carries a negative charge.

A specialized form of IEC is ion chromatography (IC), which is the name applied to the analysis of inorganic anions, cations, and low molecular weight, water-soluble organic acids and bases. Although any HPLC technique used to separate the above species can be termed ion chromatography, in general IC involves the use of ion-exchange columns and a conductivity detector. Ion chromatography itself can be subclassified. Suppressed IC involves the use of a membrane device, known as a suppressor, between the column and the detector to lower the response of the eluent and thereby enhance the signal from the solute; nonsuppressed or "single-column" IC does not contain a suppressor. Figure 2.10 shows the separation of anions in a carbonated apple juice using suppressed IC. Detailed information on the theory and applications of the analysis of ionic species can be obtained from the text by Haddad and Jackson.[25]

2.5.1 Mechanism of Retention

As an ionic solute passes through the column, it distributes itself between the mobile phase and the stationary phase by exchanging with the

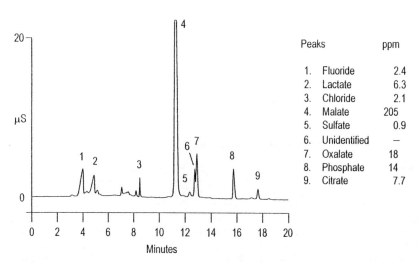

Figure 2.10 Separation of anions in a carbonated apple juice using suppressed ion chromatography. Chromatography conditions: column, AS11 with AG11 guard; detector, CD20 conductivity detector with the ASRS self-regenerating suppressor in the recycle mode. (Courtesy of Dionex Corporation.)

counterions associated with the stationary phase, as shown in Figure 2.11. As the electroneutrality of the solution must be maintained during the ion-exchange process, the exchange is stoichiometric: a single monovalent solute ion displaces a single monovalent counterion. Separation occurs as a consequence of differences in the size, charge density, and structure of the different ionic solutes.

The equilibrium constant for the exchange reaction is known as the selectivity coefficient and provides a rough means for predicting the elution order of the various ions. In general, the selectivity coefficients increase with increasing polarizing power of the solute ions. Thus, ions with high charge and a small radius of hydration should have the greatest affinity for the ion exchanger. Although this tends to be true, occasionally ion exchange is not the only operative mechanism of retention. For example, the solutes could be adsorbed onto the surface of the ion-exchange matrix. In these cases, the elution order is less predictable.

Selectivity coefficients for the interaction of cations with a strong acid cation exchanger generally fall in the order[46] $Pu^{4+} \gg La^{3+} > Ce^{3+} > Pr^{3+} > Eu^{3+} > Y^{3+} > Sc^{3+} > Al^{3+} \gg Ba^{2+} > Pb^{2+} > Sr^{2+} > Ca^{2+} > Ni^{2+} > Cd^{2+} > Cu^{2+} > Co^{2+} > Zn^{2+} > Mg^{2+} > UO_2^{2+} \gg Tl^+ > Ag^+ > Cs^+ > Rb^+ > K^+ > NH_4^+ > Na^+ > H^+ > Li^+$. This series indicates that an eluent made up using a potassium salt will be a stronger eluent than one using a sodium salt, if ion exchange is the only mechanism in operation. Similarly a series exists for the interaction of anions on strong base anion exchangers: citrate > salicylate > $ClO_4^- > SCN^- > I^- > S_2O_3^{2-} > WO_4^{2-} > MO_4^{2-} > CrO_3^{2-} > C_2O_4^{2-} > SO_4^{2-} > SO_3^{2-} > HPO_4^{2-} > NO_3^- > Br^- > NO_2^- > CN^- > Cl^- > HCO_3^- > H_2PO_4^- > CH_3COO^- > IO_3^- > HCOO^- > BrO_3^- > ClO_3^- > F^- > OH^-$.

2.5.2 Stationary Phases for Ion-Exchange Chromatography

Ion exchangers are characterized both by the type of support and by the functional group providing the charge. Functionalized silica and synthetic polymeric resins are the most common supports, although some inorganic materials are sometimes used. Synthetic polymeric resins are typically styrene–divinylbenzene or methacrylic acid–divinylbenzene copolymers treated with an appropriate reagent to produce the desired functional group. The major drawback to the use of a silica support is the pH limitation imposed by the instability of silica at high and low pH (i.e., above pH 8 and below pH 2). Synthetic polymers may be used over a much wider pH range and are often used for the analysis of carbohydrates at pH 12 or above. Synthetic polymers, however, may suffer from a degree of swelling when in contact with aqueous mobile phases, depending on the composition of the eluent.

Another important difference between silica-based columns and poly-

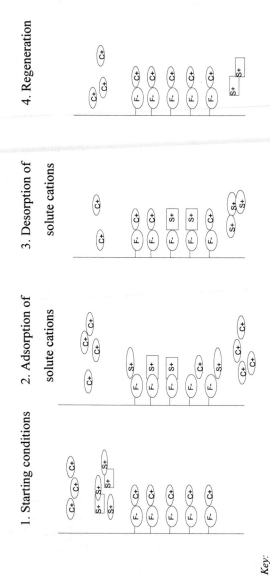

Key:

F^- = Fixed anion, C^+ = counter-ion, S^+ = solute cation

Anions in solution required for electroneutrality, are not shown for reasons of clarity.

Figure 2.11. Ion-exchange chromatography. For clarity, anions in solution required for electroneutrality are not shown. F^-, fixed anion; C^+, counterion; S^+, solute cation.

meric resins is the ion-exchange capacity. The ion-exchange capacity is determined by the number of functional groups per unit weight of resin and is usually measured as milliequivalents of charge (meq) per gram of dry resin. The ion-exchange capacity of polymeric resins can be as high as 3–5 meq/g, although much less is required for IC; silica-based resins typically have much lower capacities.

The stationary phase or ion exchanger is classified as an anion-exchange material when the functional group, or fixed ion, carries a positive charge and as a cation exchanger when it carries a negative charge. Ion exchangers are further subdivided into strong acid or base and weak acid or base types, as shown in Table 2.5. Strong ion exchangers retain the charge on the fixed ion over a wide pH range, whereas weak ones are ionized only over a much narrower pH range.[25] Most separations are performed on strong acid cation exchangers or strong base anion exchangers.

2.5.3 Mobile Phases for Ion-Exchange Chromatography

Mobile phases, or eluents, in IEC are aqueous solutions of a salt or mixture of salts, often with a small percentage of an organic solvent added. The salt mixture may be a buffer, or a buffer may be added if required. The main component of the eluent is the competing ion that causes the solute ions to be eluted.

In nonsuppressed IC, eluent competing ions of low limiting equivalent ionic conductance,[47] such as carboxylate, are required. In suppressed IC, the mechanism of suppression dictates the choice of an eluent. In the case

Table 2.5 Functional Groups on Typical Synthetic Ion-Exchange Materials[a]

Type	Functional group	Classification
Cation exchangers		
Sulfonic acid	$-SO_3^-H^+$	Strong
Carboxylic acid	$-COO^-H^+$	Weak
Phosphonic acid	$-PO_3H^-H^+$	Weak
Phosphinic acid	$-PO_2H^-H^+$	Weak
Phenolic	$-O^-H^+$	Weak
Arsonic acid	$-AsO_3H^-H^+$	Weak
Selenonic acid	$-SeO_3^-H^+$	Weak
Anion exchangers		
Quaternary amine	$-N(CH_3)_3^+OH^-$	Strong
Quaternary amine	$-N(CH_3)_2(C_2H_5OH)^+OH^-$	Strong
Tertiary amine	$-NH(CH_3)_2^+OH^-$	Weak
Secondary amine	$-NH_2(CH_3)^+OH^-$	Weak
Primary amine	$-NH_3^+OH^-$	Weak

[a] Adapted from Ref. 25 with permission.

of suppressed anion-exchange separations, cations in the eluent are replaced by hydronium ions from the anion suppressor. These hydronium ions react with the eluent anion to form an undissociated weak acid (e.g., bicarbonate forming carbonic acid), thereby reducing the conductivity of the eluent. In suppressed cation-exchange chromatography anions from the cation suppressor are replaced by hydroxide ions which react with the eluent cations to form an undissociated weak base (e.g., with hydrogen ions to form water). In suppressed IC, therefore, eluent competing ions that can be easily neutralized in an acid–base reaction, such as carbonate or mineral acids, are used.

Selectivity in the separation of ionic solutes may be varied by changing either the pH of the mobile phase or the nature or concentration of the displacing ions. The pH of the eluent affects not only the ionic form in which the eluent exists but also the form of the solute and the functional group on the ion-exchange resin. The nature and concentration of the displacing ion will determine the ease with which solute ions are displaced; the more concentrated the competing ion in the eluent, the more effectively it will displace solute ions from the stationary phase.

2.5.4 Practical Considerations

The relative affinity of an anion exchanger for different ionic solutes is a function of both the type of ion exchanger and the conditions under which it is used. Variables to consider when predicting the elution order of a series of solutes include the charge on the solute ion, the solvated size and polarizability of the solute, the degree of cross-linking of the ion-exchange resin, the ion-exchange capacity and functional group of the resin, and the degree to which the solute ion interacts with the ion-exchange matrix.[25]

There are significant differences between the behavior of small molecules and large ones during ion-exchange chromatography.[48] A relatively small change in the ionic strength of the mobile phase can cause a greater change in the elution volume of biological macromolecules relative to small molecules.[49] Column length plays a minimal role in resolution of proteins, with almost equivalent resolution being obtained with 5- or 25-cm columns.[50] However, the three-dimensional structure of a macromolecule can have a large influence on retention.[48]

2.5.5 Applications for Ion-Exchange Chromatography

Any species that can acquire a charge can be separated by ion-exchange chromatography. Some of the more important applications include analyses of amino acids, nucleotides, carbohydrates, and proteins. Carbohydrate analysis is an interesting example, as most carbohydrates are uncharged

2.5 Ion-Exchange Chromatography

below about pH 12. Therefore, an eluent such as sodium hydroxide must be chosen, which will keep the analytes charged, and a polymer-based stationary phase must be employed, as silica is unstable above about pH 8. Figure 2.12 shows the separation of myoglobin, cytochrome c, and ovalbumin on an anion-exchange resin, using a sodium chloride gradient in Bis–Tris.[51] The column contains a macroporous, hydrophilic polymer with quaternary nitrogen groups attached to the surface. As a result of the polymeric nature of the column, it is possible to use this column over a much broader range of pH values than would be possible with a silica-based column.

Ion chromatography is the method of choice for the analysis of trace and ultratrace levels of inorganic anions. It is also important for the simultaneous separation of inorganic anions and organic acids. Although IC is

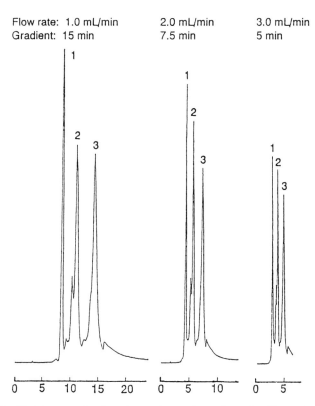

Figure 2.12 Protein separation by anion-exchange chromatography. Chromatography conditions: column, Hydrophase HP-SAX; eluent A, 5 mM Bis–Tris, pH 8.35; eluent B, A plus 0.5 M NaCl; gradient, 0–50% B over the times indicated; temperature, ambient; detection, UV absorbance at 280 nm; injection, 100 ml. Peaks: 1, myoglobin; 2, cytochrome c; 3, ovalbumin. (Reprinted from Ref. 51 with permission.)

frequently used for the analysis of inorganic cations, this is of less importance owing to the availability of other techniques.

2.6 Size-Exclusion Chromatography

Size-exclusion chromatography (SEC) is a convenient and highly predictable method for separating simple mixtures whose components are sufficiently different in molecular weight. For small molecules, a size difference of more than about 10% is required for acceptable resolution; for macromolecules a twofold difference in molecular weight is necessary.[51] Size-exclusion chromatography can be used to indicate the complexity of a sample mixture and to provide approximate molecular weight values for the components. It is an easy technique to understand, and SEC can be applied to the separation of delicate biomacromolecules as well as to the separation of synthetic organic polymers.

In SEC species in solution are separated on the basis of their molecular size, which in turn is related to the logarithm of the molecular weight. There are two modes of size-exclusion chromatography: gel-filtration chromatography (GFC) and gel-permeation chromatography (GPC). In GFC, aqueous mobile phases and hydrophilic packings are used to separate and identify biological macromolecules. On the other hand, GPC is usually performed using hydrophobic stationary phases and organic mobile phases to obtain molecular weight distribution information on polymers. Because the solutes are all eluted within a small retention volume, peaks in SEC are generally narrow, thereby enhancing sensitivity and allowing the use of relatively insensitive detection methods such as refractive index (RI) detection. Refractive index detection is particularly suitable for GPC, as many polymers have no chromophores or other detectable properties. Further practical information can be obtained from Yau *et al.*[52]

2.6.1 Mechanism of Retention

The stationary phase in SEC is a highly porous substrate whose pores are penetrated best by small solute molecules. Because larger solute molecules are unable to enter as deeply into the pores, they will travel further down the column in the same time. The largest molecules, which are totally excluded from the pores, are eluted first from the column. Because the solvent molecules are usually the smallest, they are normally the last to be eluted. The rest of the solute molecules are eluted between these two extremes, at a time dependent on their ability to penetrate into the pores. In SEC, therefore, unlike other chromatographic methods, the entire sample often is eluted before the solvent dead time peak, t_0, as shown in Figure 2.13.[53]

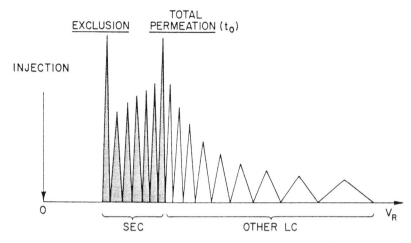

Figure 2.13 Characteristics of column dispersion and peak capacity in SEC and other LC methods. (Reprinted from Ref. 53 with permission.)

2.6.2 Stationary Phases for Size-Exclusion Chromatography

There are two classes of stationary phases in SEC, one type for GFC and the other for GPC. Stationary phases for GFC are hydrophilic and include polydextrans, polyvinyl alcohol gel, and silica gel; those for GPC are hydrophobic, typically cross-linked, rigid polystyrene–divinylbenzene gels. Generally, columns of 15 to 50 cm length are used, packed with 7- to 10-μm particles and with an internal diameter between 0.6 and 0.8 cm. In SEC, unlike in other chromatographic modes, the stationary phase is the primary factor controlling retention.

When setting up for a size-exclusion separation, the primary decision concerning the stationary phase is the pore size of the packing.[52] Columns for SEC are labeled according to pore size or exclusion limit, and, for any given column, solute retention can be represented by a graph of molecular size versus retention volume, V_R, as shown in Figure 2.14a.[54] The point at which the linear curves begin to rise sharply is the exclusion limit; larger molecules cannot enter the pores of the packing material and are eluted together as the first peak. The point at which the linear curves begin to fall sharply is known as the permeation limit; smaller molecules all have equal access to the pores and are eluted together as the last peak. The intermediate molecular weight range, falling on the linear portion of the curves, is the fractionation range and represents the useful SEC separation range for a given packing.

Small molecules (molecular weight < 5000) are separated using columns with the smallest pore sizes (60–100 Å). A column containing a

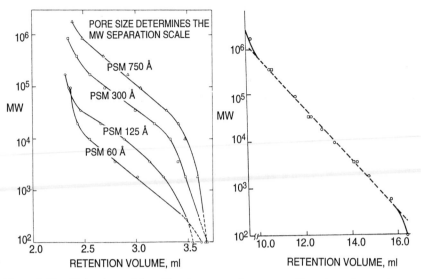

Figure 2.14 Molecular weight calibration curves for porous silica microsphere columns. Chromatography conditions: column, 10 × 0.78 cm; mobile phase, tetrahydrofuran, 22°C; flow rate, 2.5 ml/min; detection, UV absorbance at 254 nm; sample, 25 μl solutions of polystyrene standards. (a) Four individual columns showing four different calibration ranges; (b) four columns in series with two distinct pore sizes (60, 60, 750, 750Å), providing a single calibration curve with a broader molecular weight range than the individual columns. (Adapted from Ref. 54 with permission.)

packing material with a single nominal pore size is capable of separating a molecular weight range of about two orders of magnitude. By placing two columns of the same pore size in series, peak resolution can be improved; by placing columns with different pore sizes in series, the range of molecular weights that can be separated may be extended. Figure 2.14b shows a broad-range linear calibration curve obtained by placing two 60-Å columns of the type shown in Figure 2.14a, with two 750-Å columns, to produce a bimodal set. Columns for a bimodal set are chosen so that the two pore sizes differ by about one order of magnitude. In this way, the molecular weight ranges of the individual calibration curves are closely adjacent but not overlapping.

2.6.3 Mobile Phases for Size-Exclusion Chromatography

Mobile phases for SEC fall into two broad categories: aqueous buffers for GFC and organic solvents for GPC. In SEC, the mobile phase is selected not to control selectivity but for its ability to dissolve the sample. In addition, the mobile phase should have a low viscosity and be compatible with the detector and column packing. For example, polar solvents such as methanol and ethanol should not be used with polystyrene packings since they cause excessive shrinkage which will result in permanent damage to the column; instead, solvents such as toluene or chloroform should be chosen. Silica gels can be used with a wide range of solvents, including water, but they are limited to an operating range of pH 2–8.

It is also important that the mobile phase be chosen to prevent interaction of the sample components with the surface of the packing by adsorption or other unwanted effects. Non-size-exclusion effects in GFC, such as those shown in Table 2.6, can usually be avoided by selecting proper combinations of stationary and mobile phases.[55] Similarly, in GPC solvents that reduce these interactions, such as toluene or tetrahydrofuran, are commonly used. When these solvents cannot be used, salts such as lithium bromide may be added.

2.6.4 Practical Considerations

The primary factors affecting resolution in SEC include the column volume, particle size and pore size distribution, the flow rate, and the conformation of the solute; temperature and solvent viscosity, however, should also be considered. As indicated in section 2.6.2 above, one of the most common ways to increase resolution is to place two identical columns in series, thereby increasing the column volume. The major drawback to the use of columns in series, however, is the increased analysis times that result. One way to circumvent that problem is to consider the particle size of the packing material; because there is less band broadening for small packing particles, short columns containing small particle packings can be used to achieve fast SEC separations without sacrificing resolution.[52]

Table 2.6 Summary of Methods to Reduce or Eliminate Unwanted Chromatographic Effects in Gel-Filtration Chromatography[a]

Problem	Solution
Ion exchange	Increase ionic strength; adjust pH
Ion exclusion	Increase ionic strength
Ion inclusion	Increase ionic strength
Electrostatic effects of polyelectrolytes	Increase ionic strength
Adsorption	
Coulombic	Increase ionic strength; adjust pH
Hydrogen bonding	Add urea or guanidine hydrochloride
Hydrophobic	Add organic modifier; lower ionic strength
Sample viscosity effects	Increase ionic strength; lower sample concentration

[a] Reprinted from Ref. 55 with permission.

The flow rate has a significant effect on column efficiency; an increase in flow rate results in decreased efficiency. The effect is much greater for higher molecular weight materials than for lower molecular weight components. Because of the effect, SEC columns are normally operated at low flow rates. Temperature is usually applied in SEC to aid in the dissolution of the sample or to help decrease the solvent viscosity. Unlike the case for other LC methods, increasing the temperature is not an effective means to alter selectivity. The sample viscosity should be no greater than twice that of the mobile phase,[1] or peak broadening and longer elution times can result. Both of those effects will result in erroneous molecular weight assignments; thus, whenever possible, the sample should be dissolved in the mobile phase.

2.6.5 Applications for Size-Exclusion Chromatography

Because SEC is a gentle technique, rarely resulting in loss of sample or reaction, it has become a popular choice for the separation of biologically active molecules. Each solute is retained as a relatively narrow band, which facilitates solute detection with detectors of only moderate sensitivity. Examples of analytes that can be analyzed by SEC are listed in Table 2.7. Possible stationary phases and mobile phases are also provided in Table 2.7.

One of the major applications of SEC is polymer characterization. As many of the properties that characterize a polymer, including hardness, brittleness, and tensile strength, are related to the molecular weight distribution, GPC can be used to identify subtle differences between polymer materials. The GPC technique can also be used as an alternative to reversed-phase LC for samples, such as creams, ointments, and lotions. These samples contain high levels of hydrophobic materials, such as triglycerides, which could bind strongly to the reversed-phase column and cause column fouling.

Figure 2.15 shows chromatograms of octylphenoxy oligo(ethylene glycol)s which are nonionic surfactants that contain the number of ethylene oxide (EO) units indicated on the peaks in Figure 2.15.[56] Two columns of differing hydrophobicity were compared, and the octadecylsilane (ODS) column was found to be the more hydrophobic.

2.7 Affinity Chromatography

Affinity chromatography may be applied to the separation of any biological entity that is capable of forming a dissociable complex with another species, for example, proteins, nucleic acids, and even whole cells. It is based on the ability of biological macromolecules to recognize and bind to other molecules, often in a highly specific manner. The separation process takes advantage of the "lock-and-key" mechanism that is prevalent in biological systems. The strategy is completely reversible: not only can immo-

2.7 Affinity Chromatography

Table 2.7 Solvents and Packings Used in Size-Exclusion Chromatography[a]

Polymer types	Typical solvent system	Typical column packings[b]	Supplier
Proteins, polypeptides	Aqueous buffers	Zorbax Bio Series GF	Du Pont
Biopolymers, viruses, DNA, RNA	Aqueous buffers	TSK-G-PW	Toya Soda
		Superose	Pharmacia
Cellulose derivatives, polyvinyl alcohol, polysaccharides	Aqueous buffer, salts	TSK-G-SW	Toya Soda
Many polar noncrystalline synthetic polymers, some crystalline polymers, small molecules	Tetrahydrofuran	Ultra-Styragel	Waters
Nonpolar, noncrystalline synthetic polymers, hydrocarbon polymers, low molecular weight polymers	Toluene (benzene[c] or chloroform[c]	LiChrospher Si	Merck
Polar crystalline polymers (e.g., polyamides and polyesters)	m-Cresol (hot) or hexafluoroisopropanol (cold)	PL gel	Polymer Labs
Nonpolar crystalline polymers (e.g., polyethylene and stereoregular polyhydrocarbons)	1,2,4-Trichlorobenzene (hot) or 1,2-dichlorobenzene (hot)	Zorbax PSM (bimodal)	Du Pont

[a] Reprinted from Ref. 52 with permission.
[b] These examples merely illustrate commercially available packings. Each packing often can be used for many polymer types.
[c] Suspected carcinogen.

bilized antigens be used to purify antibodies, for example, but antibodies can be used to purify antigens. The texts by Lowe[57] and Lowe and Dean[58] provide practical information on the use of affinity chromatography.

2.7.1 Mechanism of Retention

The basic principle of affinity chromatography is shown in Figure 2.16.[59] An affinity ligand (biological entity) that is specific to only one type of biological molecule is covalently bonded to an inert support material. A sample containing a mixture of biomolecules is then applied to the head of the column. As the sample is washed through the column, only those species that are complementary to the affinity ligand molecules are adsorbed; the other components are eluted without retention. The adsorbed

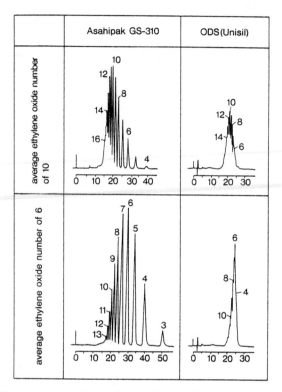

Figure 2.15 Chromatograms of octylphenoxy oligo(ethylene glycol)s on an Asahipak GS-310 column (500 × 7.6 mm I.D.) using water/acetonitrile (60:40) as the eluent and on a 5-μm Unisil Pack ODS column using water/acetonitrile (45:55) as the eluent. Other conditions: detection, UV absorbance at 280 nm; flow rate, 1 ml/min; temperature, 30°C. (Reprinted from Ref. 56 with permission.)

species are eluted after the rest of the sample by changing the composition of the mobile phase. A typical chromatogram has a nonretained peak containing several components, followed by a retained peak that ideally contains only the one, biospecifically adsorbed enzyme.[60]

2.7.2 Stationary Phases for Affinity Chromatography

The basis for selectivity in affinity chromatography is the use of immobilized biochemicals, known as affinity ligands, that are covalently attached to a support matrix, as illustrated in Figure 2.17. The primary criteria that govern the suitability of a support matrix for affinity chromatography include (1) the mechanical and flow properties of the matrix, (2) the ease of covalent coupling of the ligand to the matrix, and (3) the stability of the bonds linking the ligand to the matrix.[61] The best resolution in affinity chromatography is obtained if the matrix particles are uniform in size and shape and are as small as possible; however, rigid silica or polymeric beads, 5 or 10 µm in diameter used in normal-phase HPLC are suitable. Because the large biomolecules separated by affinity chromatography tend to diffuse slowly, the low flow rates used with low-pressure chromatography are often the conditions of choice.

Support materials for low-pressure affinity chromatography include agarose (cross-linked with epichlorohydrin), cellulose, dextran, silica, and polyacrylamide;[62] in HPLC a rigid, highly porous, hydrophilic polymer is typically used. Large pore sizes are necessary, as either the analyte or the affinity ligand are macromolecules. To provide unhindered access of

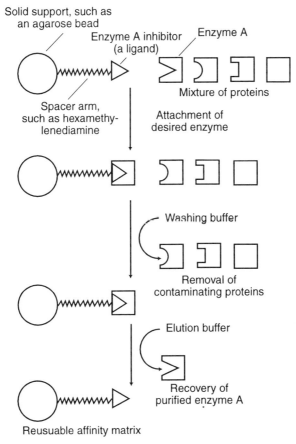

Figure 2.16 Basic principles of affinity chromatography. (Reprinted from Ref. 59 with permission.)

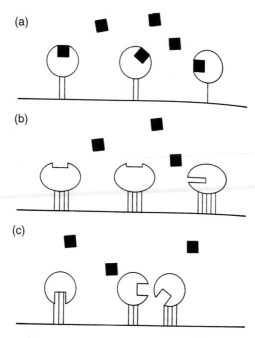

Figure 2.17 Effects of immobilization of a macromolecule. (a) Ideal case, (b) altered three-dimensional structure, and (c) improper orientation or spacing. (Reprinted from Ref. 60 with permission.)

macromolecules, pore sizes in the range of 100–400 nm are necessary. In addition, the support matrix should be stable to organic solvents as well as aqueous phases to allow greater flexibility in the selection of conditions for coupling and also for elution. A review by Chaiken et al.[63] on high-performance affinity chromatography and its use as an analytical tool can be found in *High Performance Liquid Chromatography*.

Affinity ligands can be antibodies, enzyme inhibitors, or any molecule that binds reversibly and bioselectively to the complementary analyte molecules in the sample. Reactive dyes, for example, can be used to separate most proteins, including nucleotide-binding proteins, whereas protein A and protein G can be used to bind with immunoglobulins. Hormones and drugs are often used to bind to receptors in a sample, and lectins can be used to bind with glycoproteins and cells. The ligands are usually attached or immobilized onto the support by covalent bond formation. Since many of the ligands are large, fragile molecules whose three-dimensional structure and orientation are important for activity, care must be taken during the immobilization stage.[59] When small ligands are immobilized, a spacer arm, which is generally neutral and hydrophilic in nature, is often employed between the support matrix and the affinity ligand, as shown in Figure 2.17, to enable the ligand to reach the binding site of the analyte macromolecule.

2.7.3 Mobile Phases for Affinity Chromatography

In affinity chromatography, the mobile phase is a buffer. Typically, the sample is dissolved in the buffer, and thus buffers such as potassium phosphate or (2-hydroxyethyl)-1-piperazineethanesulfonic acid (HEPES) are used. Once the unretained material has been washed from the column, the analyte must be eluted; there are two approaches to the removal of the analyte: nonspecific and biospecific elution.

Nonspecific elution involves denaturation, usually reversibly, of either the analyte or the affinity ligand by means of a gradient of pH, chaotropic agents, organic solvents, or ionic strength; if the analyte or the ligand is no longer in the correct configuration for binding, the analyte will be eluted. Biospecific elution, on the other hand, involves the inclusion of a mobile phase modifier, known as an inhibitor, that competes either with the ligand for sites on the analyte or with the analyte for sites on the ligand. The inhibitor is a free ligand similar, or identical, to the immobilized affinity ligand or the analyte. Biospecific elution is most commonly used when a low molecular weight inhibitor is available.[59] In either case, the column must be regenerated in the initial mobile phase before the next run. Figure 2.18 shows the biospecific elution of wheat germ lectin using N-acetyl-D-glucosamine as the ligand and the inhibitor.[64] The epoxy packing was able to bind covalently N-acetyl-D-glucosamine at pH 11 through one of the hydroxyl groups on the sugar.[64] A step gradient was employed to elute the protein, from 100% A (70 mM sodium phosphate, pH 7, with 150 mM sodium chloride) to 100% B (buffer A plus 10 mg/ml N-acetyl-D glucosamine) after 7 min.

2.7.4 Practical Considerations

Successful use of affinity chromatography often depends on selection of the proper chemical reactions for immobilization of the affinity ligand. Because the presence of residual ionic charges on the derivatized support matrix may produce unexpected results during purification, the cyanogen bromide method is typically the method of choice for the attachment of ligands to gels such as agarose, Sepharose, methylcellulose, or acrylamide.[65]

One of the major advantages of working with affinity chromatography is that one can concentrate very dilute solutions while simultaneously stabilizing the adsorbed protein on the column. The other major advantage is the vast range of potential ligands available; however, affinity chromatography is a comparatively expensive technique. Disadvantages include the potential for leakage of the affinity ligand from the column and the fact

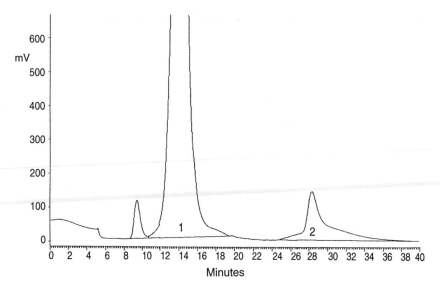

Figure 2.18 Wheat germ lectin isolation using N-acetyl-D-glucosamine as the ligand. Chromatography conditions: column, Waters AP1 glass column, 100 × 10 mm I.D.; flow rate, 1 ml/min; detection, UV absorbance at 280 nm. The sample was 1 ml of crude wheat germ preparation; 1.2 mg/ml of lectin was eluted from a crude preparation with 6.8 mg/ml total protein. (Reprinted from Ref. 64 with permission.)

that columns cannot be cleaned with aggressive chemicals such as sodium hydroxide.

2.7.5 Applications for Affinity Chromatography

Affinity chromatography may be applied to the separation from other components in a sample of any biological entity that is capable of forming a dissociable complex with another species. Examples of molecules that have been separated by affinity chromatography are listed in Table 2.8. Affinity chromatography is often used as one of a series of steps in the purification of a protein from an extract, usually, toward the end of the process, following salt fractionation and ion-exchange chromatography which are required to place the extract in a form suitable for affinity chromatography.

Quantititatively, the value of an affinity chromatography step in the purification process is assessed by the purification factor and the recovery or yield of activity. The purification factor is the ratio of the specific activity after the affinity step to that before it, whereas the recovery or activity is the percentage of the initial activity in the sample that is recovered.[61] Examples of the use of adenosine monophosphate (AMP) and related compounds, lectins and related proteins, proteinases and their protein inhibitors, and antibodies and antigens as ligands for the purification of proteins in affinity chromatography are provided in the review by Winzor and De Jersey.[61]

2.8 Chiral Separations

In "chiral" separations enantiomers are separated on the basis of stereoselectivity. Chiral chromatography has developed with the need for racemically pure drugs. Many chiral drugs have been used as their racemates[66] because of difficulties in stereoselective synthesis and purification. As only one of the enantiomers may exhibit the pharmacological effect and the other may even show side effects,[67-69] the development of analytical methods for the separation and determination of enantiomers for drug use and in biological samples, such as serum and plasma, is an area of increasing interest[70] and practical utility.

2.8.1 Mechanism of Retention

Enantiomeric compounds cannot be separated directly in normal chromatographic systems since the groups attached to the chiral atom are equally accessible to binding. If the enantiomers are combined with a chiral selector (enantiomeric reagent), however, two diastereomers are formed that may

Table 2.8 Affinity Ligands and Purified Proteins[a]

Immobilized ligand	Purified proteins
Di- and trivalent metal ions	Proteins with abundance of His, Trp, and Cys residues
Lectins	Glycoproteins, cells
Carbohydrates	Lectins
Reactive dyes	Most proteins including nucleotide-binding proteins
Nucleic acids	Exo- and endonucleases, polymerases and other nucleic acid-binding proteins
Amino acids (e.g., Lys, Arg)	Proteases
Nucleotides, cofactors, substrates, and inhibitors	Enzymes
Protein A, protein G	Immunoglobulins
Hormones, drugs	Receptors
Antibodies	Antigens
Antigens	Antibodies

[a] Reprinted from Ref. 62 with permission.

Table 2.9 Separation of Enantiomeric Compounds[a]

Principle	Chiral selector	Binding of substrate	Solid phase
Diastereomeric derivative	Precolumn	Covalent	Nonchiral
Diastereomeric complex	In liquid phase	Reversible	Nonchiral
Diastereomeric complex	Bound to solid phase	Reversible	Chiral

[a] Reprinted from Ref. 71 with permission.

show differences in binding properties and therefore can be separated by chromatography. The principles for such separations are listed in Table 2.9.[71]

There are three commonly used approaches to the separation of enantiomers by HPLC. Of these, two are based on operation with conventional reversed-phase column materials.[72] One approach is to derivatize the sample prior to chromatography, leading to formation of diastereomeric products of the two enantiomers that can be separated by conventional HPLC,[73,74] whereas the other uses chiral selectors in the mobile phase.[75,76] The third approach involves the use of chiral stationary phases.

2.8.2 Stationary Phases for Chiral Chromatography

The use of chiral stationary phases (CSP) is a technique that relies on the formation of transient, temporary diastereoisomers between the sample enantiomers and the chiral stationary phase.[1] Differences in stability between the diastereoisomers are reflected in differences in retention times; the enantiomer that forms the least stable complex is eluted first. Chiral stationary phases may be classified into five different groups on the basis of mechanism of retention:[77] (1) chiral phases with cavities[78-80] (inclusion mechanism), shown in Figure 2.19, (2) chiral affinity phases,[81,82] (3) chiral phases based on multiple hydrogen bond formation,[83] (4) chiral π-donor and

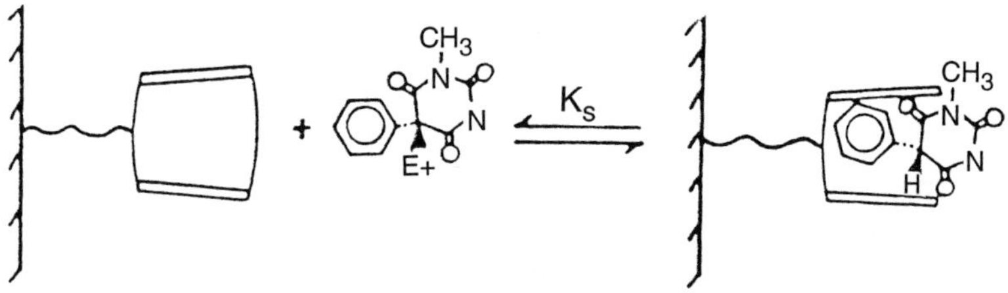

Figure 2.19 Inclusion mechanism of retention for mephobarbital in cyclodextrin. (Reprinted from Ref. 80 with permission.)

2.8 Chiral Separations

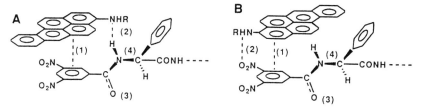

Figure 2.20 Proposed three-point interaction model between Pirkle-type chiral stationary phase and the best orientations of 3-aminobenzo[a]pyrene for maximum interaction. (Adapted from Ref. 111 with permission.)

π-acceptor phases,[84] shown in Figure 2.20, and (5) chiral ligand-exchange chromatography phases.[85] Table 2.10 lists examples of chiral selectors from each of the above categories, with possible substrates.

The most commonly encountered stationary phases that take advantage of the inclusion mechanism are the cellulose derivatives. However, cyclodextrins,[86,87] synthetic chiral polymer phases[88] such as polyacrylamide and polymethacrylamide gels, chiral crown ethers,[89] and chiral imprinted gels[90] are also frequently used. The inclusion mechanism is a popular mechanism and has been applied to the separation of a wide variety of drugs, including β-blockers, barbiturates, adrenergic drugs, antihistaminic drugs, atropine, and cocaine.[77]

With chiral affinity phases, proteins undergo enantioselective interactions with a great variety of drugs. Thus, the resolution on chiral affinity stationary phases is due to interactions of the enantiomers with proteins bonded to the solid support. Typical proteins used for chiral affinity separa-

Table 2.10 Chiral Selectors Bound to the Solid Phase[a]

Principle	Chiral selector (example)	Ref.	Mobile phase	Substrate (example)
Inclusion	Triacetylcellulose	78	Organic	Aromatic compounds
	β-Cyclodextrin	79	Polar	Amides, imides
Affinity	Albumin	81	Aqueous	Acidic compounds
	α_1-Acid glycoprotein	82	Aqueous	Amines, carboxylic acids, nonprotolytes
Multiple H-bond	Polymethacrylate	83	Organic	Aromatic compounds
π-donor/acceptor phases	(R)-N-(3,5-Dinitrobenzoyl)phenyl-glycine	84	Organic	Amides, cyclic imides, carbamates
Ligand	L-Proline amide + Cu(II)	85	Aqueous	Amino acids

[a] Adapted from Ref. 71 with permission.

tions include bovine serum albumin[91] (BSA), human serum albumin[92] (HSA), ovomucoid,[93] and avidin.[93] The primary interactions responsible for chiral recognition are hydrophobic and coulombic interactions,[93] but the hydrophobic spacer linking the protein to the solid support can significantly affect retention and separation of the enantiomers. Changes in the properties of the mobile phase, such as the type and concentration of a modifier or pH, can be used to affect the retention and separation of the enantiomers. Further information on the use of this type of chiral phase is available in Refs. 71 and 94.

The mechanism of retention on chiral phases that is based on multiple hydrogen bonding formation involves the formation of "base pairs" and triple hydrogen bonds between the solutes and the chiral stationary phase.[95] Fundamental work in this area has been done by Hara and Dobashi,[96,97] using amino acid amide and tartaric acid amide phases. In addition, N,N'-2,6-pyridinediyl bis(alkanamides) chemically bonded to silica gel have been described for the resolution of barbiturates.[95]

The pioneering work in the area of chiral π-donor and π-acceptor phases was done by Pirkle and co-workers,[98,99] and the most frequently used π-acceptor phase is still an (R)-N-(3,5-dinitrobenzoyl)phenylglycine phase, the so-called Pirkle phase. With these stationary phases, chiral recognition is based on π–π interactions, "dipole stacking" interactions, and hydrogen bonding.[100]

The ingredients of the mobile phase are chosen on the basis of the ability to interact with the stationary phase. Hexane and heptane are nonselective solvents serving only to alter the strength of the mobile phase. On the other hand, 2-propanol can hydrogen bond at the amide group, dichloromethane is polar and will interact with the carbonyl group through dipole–dipole interaction, and chloroform is a proton donor that interacts with the amide group at the carbonyl oxygen; each of those interactions is capable of reducing the strength of the solute–CSP interaction.

Mobile phases are usually binary or ternary mixtures of solvents. Selectivity is affected mostly by mobile phase composition rather than strength, and peak shape and retention are both influenced by the addition of organic modifiers.[101] Some compounds naturally have π-donor or π-acceptor groups and can be resolved directly. In many cases, however, introduction of π-donating groups by derivatization steps is necessary. Figure 2.20 shows the proposed three-point interaction of 3-aminobenzo[a]pyrene, a polycyclic aromatic hydrocarbon (PAH), with a Pirkle-type stationary phase.[111] Two possible interactions are illustrated, showing the best orientations for maximum interaction.

Chiral ligand-exchange chromatography resolves enantiomers on the basis of their ability to complex with transition metal ions, such as copper, zinc, and cadmium, as illustrated by the separation of amino acid racemates using copper[102] (Fig. 2.21). The principle of exchange is similar to that

2.8 Chiral Separations

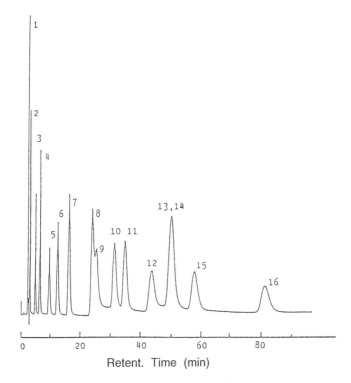

Figure 2.21 Resolution of eight amino acid racemates. Chromatography conditions: column, MCI GEL CRS10W; eluent, 0.5 mM copper(II) sulfate solution; flow rate, 1.0 ml/min; column temperature, 32°C; detection, UV absorbance at 254 nm. Peaks: 1, D-Ala; 2, L-Ala; 3, D-Pro; 4, D-Val; 5, L-Pro; 6, L-Val; 7, D-Leu; 8, D-Nle; 9, D-Tyr; 10, L-Leu; 11, D-Eth; 12, L-Tyr; 13, L-Nle; 14, D Phe; 15, L-Eth; 16, L-Phe. (Reprinted from Ref. 103 with permission.)

described for ion-exchange, but chemically modified silica gels are employed, consisting of amino acids bonded to silica gel by means of 3-glycidoxytrimethoxysilane.[103] The analytes are removed from the stationary phase by the use of a transition metal in the mobile phase. These phases show high enantioselectivity for underivatized amino acids, several amino acid derivatives, dipeptides, 3-hydroxy-L-tyrosine (Dopa), and thyroid hormones, among others.[77] However, commercially available ligand-exchange phases have shortcomings, namely, that the number of amino acids separated on any given chiral column is limited and that the column temperature must be elevated in order to increase the degree of resolution.

2.8.3 Mobile Phases for Chiral Chromatography

Because chiral stationary phases are often expensive, and sometimes have limited lifetimes, interest continues in the search for chiral mobile phase additives for use with conventional reversed-phase column materials. The combination of a large number of achiral support phases with the many potential additives as chiral selectors can lead to a wide variety of options for the separation of enantiomers. Additives can include amino acid derivatives,[104] certain proteins,[105] and inclusion complexing agents such as cyclodextrins,[72] and crown ethers.[106] Figures 2.22 and 2.23 show chromatograms obtained by the use of the inclusion mechanism[107] and ligand exchange,[108] respectively, using chiral additives in the mobile phase.

The use of chiral derivatizing agents, precolumn, to produce diastereomeric pairs is dependent on the availability of high purity derivatizing agents. Unfortunately, the rates of reaction of the enantiomers with the derivatizing agents are often different from one another, which results in formation of diastereoisomers in proportions that differ from those of the enantiomers present in the racemate. Derivatizing agents were originally used for the resolution of drugs and their metabolites in biological matrices,[109] but are now widely used in a number of fields.

2.9 Summary of Major Concepts

1. There are five major chromatographic modes that can be applied to the analysis of solutes in solution: normal phase, reversed phase, ion exchange, size exclusion, and affinity. In addition, a variety of submodes exist, such as hydrophobic interaction, chiral separations, ion suppression, and ion pairing.

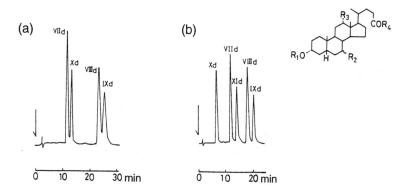

Figure 2.22 Separation of derivatives of glycine-conjugated bile acid 3-sulfates with 1-bromoacetylpyrene. Mobile phase conditions: (a) acetonitrile/0.5% KH_2PO_4 (pH 4.0); (b) same as (a) with 2 mM Me-β-CD (3:4, v/v). (Reprinted from Ref. 107 with permission.)

2.9 Summary of Major Concepts

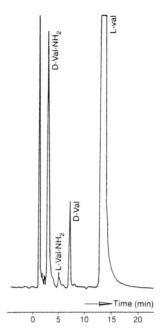

Figure 2.23 Chromatogram of sample from L-valine synthesis. Chromatography conditions: column, Nucleosil 120-5C_{18}; mobile phase, 2 mM copper(II) acetate/4 mM L-DPA/1 mM TEA (pH 5.1). (Reprinted from Ref. 108 with permission.)

2. There are a number of applications of HPLC:

normal-phase chromatography is used for the separation of neutral species on the basis of polarity,
reversed-phase chromatography is used for the separation of neutral species on the basis of hydrophobicity,
ion-exchange chromatography is used for the separation of ionic solutes on the basis of charge,
size-exclusion chromatography is used for the separation of molecules on the basis of differences in molecular size, and
affinity chromatography is used for the separation of biomolecules on the basis of the lock-and-key mechanism prevalent in biological systems.

3. The strengths of various modes are summarized in Table 2.11.
4. The limitations of various modes are summarized in Table 2.12.
5. Applications of various modes are summarized in Table 2.13.

Table 2.11 Strengths of Various Chromatographic Modes[a]

Ion exchange	Hydrophobic interaction	Affinity	Reversed phase	Gel filtration
Concentrates sample	Concentrates sample	Concentrates sample	Concentrates sample	
Good recovery of activity	Good recovery of activity	Good recovery of activity		Good recovery of activity
Moderate to high resolution	Moderate resolution	Outstanding resolution	Good resolution for lower molecular weight species	Reliable medium to high resolution
High capacity	High capacity	High capacity	Moderate capacity	
Column can be cleaned with NaOH	Column can be cleaned with NaOH	Significant purification in a single step	Compatible with detergents	Can be used in buffer exchange
Widespread applicability		Superb selectivity	Simple technique	Rapid

[a] Adapted from Ref. 110 with permission.

Table 2.12 Limitations of Various Chromatographic Modes[a]

Ion exchange	Hydrophobic interaction	Affinity	Reversed phase	Gel filtration
Fractions of biological interest may have to be desalted prior to next purification	Fractions of biological interest may have to be desalted prior to next purification		Limited applicability to high molecular weight proteins	Moderate resolution
		Column cannot be cleaned with aggressive chemicals	Column cannot be cleaned with aggressive chemicals	Nonconcentrating
		Stationary phase can be expensive	Organic mobile phase may denature protein	Moderate capacity

[a] Adapted from Ref. 110 with permission.

Table 2.13 Applications of Various Chromatographic Modes[a]

Ion exchange	Hydrophobic interaction	Affinity	Reversed phase	Gel filtration
Protein isolation and purification	Protein isolation and purification	Protein isolation and purification	Protein isolation and purification	Protein isolation and purification
Peptide analysis and purification	Peptide analysis and purification		Peptide analysis and purification	
Nucleic acid analysis			Nucleic acid analysis	Nucleic acid analysis
Complex carbohydrates				Complex carbohydrates
				Polymers
Sugars			Sugars	Sugars
			Fatty acids	
			Triglycerides and cholesterol	Triglycerides and cholesterol
			Carbamates	
Organic acids			Organic acids	
Inorganic anions and cations			Inorganic anions and cations	
Amines and amino acids			Amines and amino acids	

[a] Adapted from Ref. 110 with permission.

References

1. Snyder, L. R., and Kirkland, J. J., "Introduction to Modern Liquid Chromatography," 2nd Ed. Wiley, New York, 1979.
2. Scott, R. P. W., *Adv. Chromatogr.* **20**, 167 (1982).
3. Robards, K., Haddad, P. R., and Jackson, P. E., "Principles and Practice of Modern Chromatographic Methods." Academic Press, San Diego, 1994.
4. Salotto, A. W., Weiser, E. L., Caffey, K. P., Carty, R. L., Racine, S. C., and Snyder, L. R., *J. Chromatogr.* **498**, 55 (1990).
5. Braithwaite, A., and Smith, F. J., "Chromatographic Methods," 4th Ed. Chapman & Hall, New York, 1990.
6. Snyder, L. R., Glach, J. L., and Kirkland, J. J., "Practical HPLC Method Development." Wiley, New York, 1988.
7. Schoenmakers, P. J., *J. Chromatogr. Libr. Ser.* **35** (1986).
8. Demchak, R. J., and MacConnell, J. G., *J. Chromatogr.* **511**, 353 (1990).
9. Timble, D. J., and Keeney, P. G., *J. Food Sci.* **42**, 1590 (1977).
10. Black, L. T., and Bagley, E. B., *Am. Oil Chem. Soc. J.* **55**, 228 (1978).
11. Meagher, R. B., and Furst, A., *J. Chromatogr.* **117**, 121 (1976).
12. Gum, E. K. Jr., and Brown, R. D., *J. Anal. Biochem.* **82**, 372 (1977).
13. Poole, C. F., and Poole, S. K., "Chromatography Today." Elsevier, New York, 1991.
14. Horvath, C., Melander, W., and Molnar, I., *J. Chromatogr.* **125**, 129–156 (1976).
15. Melander, W. R., and Horvath, C., "High Performance Liquid Chromatography, Advances and Perspectives" (C. Horvath, ed.), pp. 113–119. Academic Press, New York, 1980.
16. Horvath, C., and Melander, W. R., *Amer. Lab.* 17–36 (1978).
17. Miller, J. M., "Chromatography: Concepts and Contrasts." Wiley, New York, 1988.
18. Martire, D. E., and Boehm, R. E., *J. Phys. Chem.* **87**, 1045–1062 (1983).
19. Dill, K. A., *J. Phys. Chem.* **91**, 1980–1988 (1987).
20. Dill, K. A., Naghizadeh, J., and Marqusee, J. A., *Annu. Rev. Phys. Chem.* **39**, 425–461 (1988).
21. Dorsey, J. G., and Dill, K. A., *Chem. Rev.* **89**, 331–346 (1989).
22. Colin, H., and Guiochon, G., *J. Chromatogr.* **141**, 289–312 (1977).
23. Karger, B. L., and Giese, R. W., *Anal. Chem.* **50**, 12 (1978).
24. Bidlingmeyer, B. A., *J. Chromatogr. Sci.* **18**, 525–539 (1980).
25. Haddad, P. R., and Jackson, P. E., *J. Chromatogr. Libr.* **46**, pp. 230–232 (1990).
26. Abbott, S. R., *J. Chromatogr. Sci.* **18**, 540 (1980).
27. Dionex Technical Note 12-R, Dionex Corporation, Sunnyvale, California.
28. Kissinger, P. T., *Anal. Chem.* **49**, 883 (1977).
29. Gloor, R., and Johnson, E. L., *J. Chromatogr. Sci.* **15**, 413–423 (1977).
30. De Antonis, K. M., Brown, P. R., Yi, Z., and Maugle, P. D., *J. Chromatogr.* **632**, 91 (1993).
31. Cooke, N. H. C., and Olsen, K., *J. Chromatogr. Sci.* **18**, 512–524 (1980).
32. Karger, B. L., and Giese, R. W., *Anal. Chem.* **50**, 1048A–1073A (1978).
33. Yarmchuk, P., Weinberger, R., Hirsch, R. F., and Love, L. J. C., *J. Chromatogr.* **283**, 47–60 (1984).

34. Love, L. J. C., Habarta, J. G., and Dorsey, J. G., *Anal. Chem.* **56**, 1132A–1148A (1984).
35. Gooding, K. M., and Regnier, F. E., "HPLC of Biological Macromolecules: Methods and Applications." Dekker, New York, 1990.
36. Garnick, R. L., Solli, N. J., and Papa, P. A., *Anal. Chem.* **60**, 2546 (1988).
37. Dong, M. W., and Tran, A. D., *J. Chromatogr.* **499**, 125 (1990).
38. Benedek, K., *J. Chromatogr.* **458**, 93 (1988).
39. Schmuck, M. N., Nowlan, M. P., and Gooding, K. M., *J. Chromatogr.* **371**, 55 (1986).
40. Kato, Y., Kitamura, T., and Hashimoto, T., *J. Chromatogr.* **333**, 202 (1985).
41. Hjerten, S., Yao, K., Eriksson, K.-O., and Johansson, B., *J. Chromatogr.* **359**, 99 (1986).
42. Heinitz, M. L., Kennedy, L., Kopaciewicz, W., and Regnier, F. E., *J. Chromatogr.* **443**, 173 (1988).
43. Kato, Y., Kitamura, T., and Hashimoto, T., *J. Chromatogr.* **298**, 407 (1984).
44. Wetlaufer, D. B., and Koenigbauer, M. R., *J. Chromatogr.* **359**, 55 (1986).
45. El Rassi, Z., De Ocampo, L. F., and Bacolod, M. D., *J. Chromatogr.* **499**, 141 (1990).
46. Peters, D. G., Hayes, J. M., and Heiftje, G. M., "Chemical Separations and Measurements," p. 583. Saunders, Philadelphia, Pennsylvania, 1974.
47. CRC Handbook of Chemistry and Physics, 65th ed. p. D–167. CRC Press, Inc., Cleveland, OH (1984).
48. Barth, H. G., Barber, W. E., Lochmuller, C. H., Majors, R. E., and Regnier, F. E., *Anal. Chem.* **58**, 211R (1986).
49. Ekstroem, B., *Anal. Biochem.* **142**, 134 (1984).
50. Vanecek, G., and Regnier, F. E., *Anal. Biochem.* **109**, 345 (1980).
51. Nelson, N. F., and Kitagawa, N., *J. Liq. Chromatogr.* **13**(20), 4037 (1990).
52. Yau, W. W., Kirkland, J. J., and Bly, D. D., in "High Performance Liquid Chromatography" (P. R. Brown and R. A. Hartwick, eds.), *Chemical Analysis*, Vol. 98. pp. 277–316, Wiley, New York, 1989.
53. Yau, W. W., Kirkland, J. J., and Bly, D. D., "Modern Size-Exclusion Liquid Chromatography." Wiley, New York, 1979.
54. Yau, W. W., Ginnard C. R., and Kirkland, J. J., *J. Chromatogr.* **149**, 465 (1978).
55. Barth, H. G., *J. Chromatogr. Sci.* **18**, 409 (1980).
56. Noguchi, K., Yanagihara, Y., Kasai, M., and Katayama, B., *J. Chromatogr.* **461**, 365 (1989).
57. Lowe, C. R., An introduction to affinity chromatography, in "Laboratory Techniques in Biochemistry and Molecular Biology" (T. S. Work and E. Work, eds.), Vol. 7, Part 2. Elsevier, Amsterdam, 1979.
58. Lowe, C. R., and Dean, P. D. G., "Affinity Chromatography." Wiley, Chichester, 1974.
59. Parikh, I., and Cuatrecasas, P., *Chem. Eng. News*, **August**, 17 (1985).
60. Walters, R. R., *Anal. Chem.* **57**, 1099A (1985).
61. Winzor, D. J., and De Jersey, J., *J. Chromatogr., Biomed. Appl.* **492**, 377 (1989).
62. Jones, K., *Chromatographia* **32**, 469 (1991).
63. Chaiken, I. M., Fassina, G., and Caliceti, P., in "High Performance Liquid

Chromatography" (P. R. Brown and R. A. Hartwick, eds.), *Chemical Analysis, Vol. 98.* pp. 317–336, Wiley, New York, 1989.
64. Phillips, D. J., Bell-Alden, B., Cava, M., Grover, E. R., Mandeville, W. H., Mastico, R., Sawlivich, W., Vella, G., and Weston, A., *J. Chromatogr.* **536**, 95 (1991).
65. "Sigma Affinity Chromatography Media." Sigma Chemical Company, St. Louis, Missouri, 1989.
66. Jamali, F., *Eur. J. Drug Metab. Pharmacokinet.* **13**, 1 (1988).
67. Ariens, E. J., *Eur. J. Clin. Pharmacol.* **26**, 663 (1984).
68. Ariens, E. J., *Med. Res. Rev.* **6**, 451 (1986).
69. Ariens, E. J., *Med. Res. Rev.* **7**, 367 (1987).
70. Wozniak, T. J., Bopp, R. J., and Jensen, E. C., *J. Pharm. Biomed. Anal.* **9**, 363 (1991).
71. Hermansson, J., and Schill, G., in "High Performance Liquid Chromatography" (P. R. Brown and R. A. Hartwick, eds.), *Chemical Analysis, Vol. 98.* pp. 337–374, Wiley, New York, 1989.
72. Mitchell, P., and Clark, B. J., *Anal. Proc.* **30**, 101–103 (1993).
73. Mehvar, R., and Jamali, F., *Pharm. Res.* **5**, 53 (1988).
74. Bjorkman, S., *J. Chromatogr.* **339**, 339 (1985).
75. Szepesi, G., Gazdag, M., and Ivancsics, R., *J. Chromatogr.* **244**, 33 (1982).
76. Pettersson, C., and Josefsson, M., *Chromatographia* **21**, 321 (1986).
77. Gubitz, G., *Chromatographia* **30**, 555–564 (1990).
78. Hahli, H., and Mannschreck, A., *Angew. Chem.* **89**, 419 (1977).
79. Armstrong, D. W., and DeMond, W., *J. Chromatogr. Sci.* **22**, 411 (1984).
80. Ward, T. J., and Armstrong, D. W., *J. Liq. Chromatogr.* **9**, 407 (1986).
81. Allenmark, S., Blomgren, B., and Boren, H., *J. Chromatogr.* **237**, 473 (1982).
82. Hermansson, J., *J. Chromatogr.* **298**, 67 (1984).
83. Blaschke, G., *Angew. Chem.* **92**, 14 (1980).
84. Pirkle, W. H., and Finn, J., in "Asymmetric Synthesis" (J. D. Morrison, ed.), pp. 87–124. Academic Press, New York, 1983.
85. Davankov, V. A., Kurganov, A. A., and Bochkov, A. S., *Adv. Chromatogr.* **21**, 71 (1984).
86. Wainer, I. W., *J. Pharm. Biomed. Anal.* **7**, 1033 (1989).
87. Haginaka, J., and Wakai, J., *Anal. Chem.* **62**, 997 (1990).
88. Kinkel, J. N., Fraenkel, W., and Blaschke, G., *Kontakte*, **1**, 3 (1987).
89. Motellier, S., and Wainer, I. W., *J. Chromatogr.* **516**, 365 (1990).
90. Sellergren, B., *Chirality* **1**, 63 (1989).
91. Allenmark, S., and Andersson, S., *Chirality* **1**, 154 (1989).
92. Domenici, E., Bertucci, C., Salvadori, P., Felix, G., Cahagne, I., Motellier, S., and Wainer, I. W., *Chromatographia* **29**, 170 (1990).
93. Oda, Y., Mano, N., Asakawa, N., Yoshida, Y., Sato, T., and Nakagawa, T., *Anal. Sci.* **9**, 221 (1993).
94. Noctor, T. A. G., and Wainer, I. W., *J. Liq. Chromatogr.* **16**(4), 783 (1993).
95. Feibush, B., Figueroa, A., Charles, R., Onan, K. D., Feibush, P., and Karger, B. L., *J. Am. Chem. Soc.* **108**, 3310 (1986).
96. Hara, S., and Dobashi, A., *J. Liq. Chromatogr.* **2**, 883 (1979).
97. Dobashi, A., Dobashi, Y., and Hara, S., *J. Liq. Chromatogr.* **9**, 243 (1986).

98. Pirkle, W. H., House, D. W., and Finn, J. M., *J. Chromatogr.* **192**, 143 (1980).
99. Pirkle, W. H., Finn, J. M., Schreiner, J. L., and Hamper, B. C. J., *J. Am. Chem. Soc.* **103**, 3964 (1981).
100. Pirkle, W. H., Welch, C. J., and Hyun, M. H., *J. Org. Chem.* **48**, 5022 (1983).
101. Dhanesar, S. C., and Gisch, D. J., *J. Chromatogr.* **461**, 407 (1988).
102. Kiniwa, H., Baba, Y., Ishida, T., and Katoh, H., *J. Chromatogr.* **461**, 397 (1989).
103. Gubitz, G., Jellenz, W., and Santi, W., *J. Chromatogr.* **203**, 377 (1981).
104. Berthod, A., Heng, L. J., Beesley, T. E., Duncan, J. D., and Armstrong, D. W., *J. Pharm. Biomed. Anal.* **8**, 123 (1990).
105. Levin, S., and Grushka, E., *Anal. Chem.* **57**, 1830 (1985).
106. Gazdag, M., Szepesi, G., and Huszar, L., *J. Chromatogr.* **436**, 31 (1988).
107. Shimada, K., Komine, Y., and Mitamura, K., *J. Chromatogr.* **565**, 111 (1991).
108. Duchateau, A., Crombach, M., Aussems, M., and Bongers, J., *J. Chromatogr.* **461**, 419 (1989).
109. Bjorkman, S., *J. Chromatogr.* **414**, 465 (1987).
110. "The Waters Chromatography Handbook 1993-1994," P/N WAT056930. Waters Chromatography, Milford, Massachusetts.
111. Lai, J.-S., Hung, S. S., Unruh, L. E., Jung, H. and Fu, P. P., *J. Chromatogr.* **461**, 327 (1989).

CHAPTER 3

Instrumentation for High-Performance Liquid Chromatography

3.1 Introduction

The basic components of a high-performance liquid chromatographic system are shown in Figure 3.1. The instrument consists of (a) eluent containers for the mobile phase, (b) a pump to move the eluent and sample through the system, (c) an injection device to allow sample introduction, (d) a column(s) to provide solute separation, (e) a detector to visualize the separated components, (f) a waste container for the used solvent, and finally (g) a data collection device to assist in interpretation and storage of results.

The pump, injector, column, and detector are connected with tubing of narrow inner diameter. The inner diameter of the tubing that is used between the injector and column and also between the column and the detector must be as narrow as possible (0.010 inch or less for analytical work) to minimize band broadening. The choice of detector is based on intrinsic properties of the solute. Often more than one detector can be used to maximize sample information and confirm peak identities. For example, an absorbance detector could be placed in series with a conductivity detector for the visualization of a charged, chromophoric solute.

This chapter discusses features of the pump, injector, column, and detector. In addition, approaches to sample preparation are introduced. The data station is discussed in Chapter 7 under data handling.

3.2 Solvent Delivery Systems

The function of the solvent delivery system is to deliver the mobile phase (eluent) through the chromatograph, accurately and reproducibly. The solvent delivery system comprises the pump, check valves, flow controllers, pulse dampeners, and pressure transducers, each of which needs to be maintained to ensure reproducible flow rates. The pump must be easy to prime, reliable, and easy to maintain. It should be resistant to corrosion from the eluents employed and be capable of delivering the required flow rates with minimal holdup volume for rapid solvent changes. Delivery of the mobile phase must be pulse free to ensure minimal baseline noise from the pump.

3.2.1 Pumps

The pumping systems used in HPLC can be categorized in three different ways. The first classification is according to the eluent flow rate that the pump is capable of delivering. The second classification is according to the construction materials, and the final classification is according to the mechanism by which the pump delivers the eluent. Each of these classifications is considered below.

(i) Pump Classification According to Flow Rate

When classified in terms of flow rate, pumps may be defined as microbore, standard bore, or preparative, as illustrated in Figure 3.2. Standard-bore systems are the most commonly used pumping systems for analytical HPLC because they provide reliable operation at flow rates ranging from 100 μl/min to 10 ml/min. These versatile systems are compatible with the less demanding applications of semipreparative work (12 mm columns) as well as with microbore operation (2 mm columns).

Microbore systems are intended for use with column diameters ranging up to 2 mm. The narrow column diameter and small size of the packing material decree relatively low flow rates for the pumping system, from 1 to 250 μl/min. As the minimum head size for reciprocating pumps is around 25 μl, smooth, reliable operation at flow rates less than 10 μl/min is difficult.

Preparative chromatography is an area outside the scope of analytical

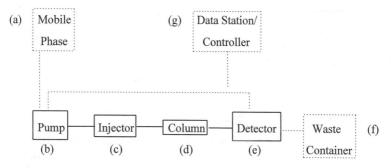

Figure 3.1 Schematic diagram of the basic modules of an HPLC system.

3.2 Solvent Delivery Systems

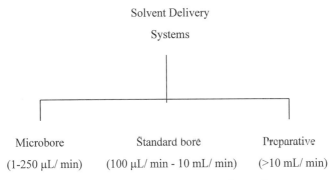

Figure 3.2 Classification of pumps according to flow rate.

HPLC covered in this book. Flow-rate ranges required for preparative work place special design constraints on instrumentation. Laboratory-scale pumps used for preparative work will typically function to 50 ml/min.

(ii) Pump Classification According to Materials of Construction

Pumps may also be classified according to the primary construction materials. As illustrated in Figure 3.3, pumps are classified as metallic or nonmetallic, depending on the material used for the eluent flow path. The most commonly used material for HPLC pumping systems is 316 stainless steel because of its mechanical strength, corrosion resistance, good thermal stability, and malleability; only a handful of HPLC solvents, such as hydrochloric acid, will cause damage to 316 stainless steel.

A material similar in most mechanical properties to 316 stainless steel is titanium, which is claimed to have a greater degree of inertness to corrosion and sample adsorption. As with steel, titanium is capable of withstanding pressures in excess of 6000 psi (41 MPa). However, the high cost of

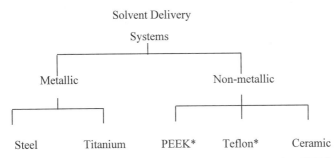

Figure 3.3 Classification of pumps according to material of construction. PEEK, Polyethylethylketone; Teflon, polytetrafluoroethylene.

titanium and its tendency to act as an ion exchanger under some circumstances are major impediments to its broad acceptance.

Some applications require the use of materials that are capable of constant exposure to high salt or extremely acidic or basic mobile phases; thus, the use of metal in the eluent flow path is precluded. In addition, many biological samples are readily degraded by contact with metals. Therefore, pumps are also constructed from nonmetallic materials, such as PEEK (polyethylethylketone), Teflon (polytetrafluoroethylene), and ceramics. Teflon has been used extensively in many "low-pressure" semipreparative HPLC applications for biological samples, but it has insufficient mechanical strength for consistent operation above 2000 psi. On the other hand, PEEK provides greater mechanical strength than Teflon, with reliable performance to 5000 psi, and it is inert to the solvents known to damage stainless steel. However, PEEK is incompatible with many of the organic solvents that are commonly used in normal-phase HPLC.

Ceramic materials, including sapphire, have been used extensively in HPLC pumps for more than 20 years as pistons and check valve components. These materials have also been used to construct heads because of their good chemical stability. The use of ceramics is limited, however, because of high cost and brittleness. Although many systems have one material as the primary construction material, the wetted surfaces of a pumping system can contain several other materials. Therefore, for material-sensitive applications, all the materials in the HPLC eluent flow path should be considered. Materials that may be encountered are polymeric materials for pump seals such as fluoropolymers, polypropylene, and Teflon; sapphire pump pistons and check valve seats; ruby check valve balls; Kalrez, KelF, or ceramic washers and spacers; polymer-based transducer components; and in older systems connections and joints made with silver solder.

(iii) Pump Classification According to Mechanism of Eluent Displacement

The third classification of pumps is according to the mechanism by which the liquid is forced through the chromatograph (Fig. 3.4). Although a wide variety of pump designs have been developed over the years, nearly all LC pumps since the 1980s are based on some variation of the reciprocating-piston pump.

A. Syringe Pumps The syringe, or positive displacement, pump remains popular for applications requiring 'pulseless' solvent delivery, such as in capillary LC (where typical flow rates are less than 100 μl/min) or in microbore HPLC connected to an interface to mass spectometry. The syringe pump is essentially a large-barrel syringe on the order of 10–50 ml, with the plunger connected to a digital stepping motor or precision screw drive. As the plunger moves forward, it drives the eluent through the chromatograph with a pulseless flow. However, the run

3.2 Solvent Delivery Systems

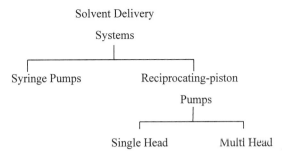

Figure 3.4 Classification of pumps according to the mechanism of eluent displacement.

time is limited by the volume of the syringe, and no flow occurs during the refill step.

B. Reciprocating-Piston Pumps The reciprocating-piston pump is the most common design in modern HPLC and is illustrated in Figure 3.5.[1] The pump head consists of two sets of moving parts: the check valves and seal–piston assembly. The cam and connecting rod transform the rotational movement of the motor into linear movement of the piston. Each stroke of the piston displaces a small volume of liquid (40–400 ml) from a chamber equipped with check valves. During the fill stroke the piston is withdrawn into the chamber. The inlet check valve rises from its seat because the incoming solvent is at a higher pressure than the pressure inside the liquid chamber of the pump head. At the same time, the outlet check valve drops down onto its seat because the pressure on the column side of the pump is also higher than that inside the pump head. Because of the two check valves, the solvent is able to enter the liquid chamber from the solvent reservoir only.

During the delivery stroke, the piston moves into the liquid chamber and pressurizes the liquid contained in it. Since the pressure inside the chamber is now greater than atmospheric pressure, the inlet check valve is forced to close. When the pressure inside the pump head exceeds the

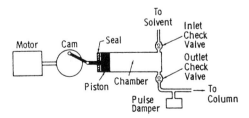

Figure 3.5 Reciprocating-piston pump head showing the check valves, piston, seal, and pulse damper. (Adapted from Ref. 1 with permission.)

pressure on the column side of the pump, the outlet check valve opens and the mobile phase flows toward the column.

3.2.2 Minimization of Pump Pulsations

(i) Cam Design

The flow delivered by a reciprocating pump is always pulsating in synchronization with the piston movement. Although pump pulsations have few adverse effects on a separation, detection of analytes at trace levels is often limited by baseline noise from pump pulsations.[2] The flow delivered by a single-piston pump is relatively pulsating so single-piston pumps are rarely used to deliver eluent; instead, they may be used in dedicated isocratic systems where a significant pulse damper volume is acceptable, for example, for sample delivery or to deliver postcolumn derivatizing solutions.

To minimize pump pulsations, pumping systems typically use a two-head design with a noncircular cam to drive the piston. The noncircular cam causes the piston to speed up during the refill stroke, thereby producing a uniform delivery over most of the cycle, regardless of the flow rate. Generally, both pistons in a dual-head reciprocating-piston pump are driven by a single motor through a common, noncircular cam, although a number of two-head designs now use separate motors for each head. In either case, the pistons are 180° out of phase. As one liquid chamber is emptying, the other is refilling; thus, the two flow profiles overlap, as illustrated in Figure 3.6, and the peaks and troughs in the total flow output are canceled.[3] Triple-

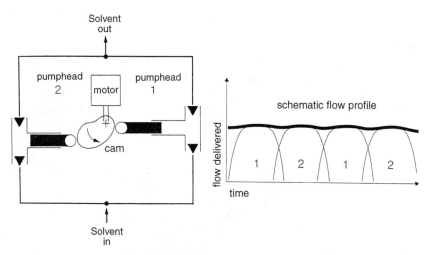

Figure 3.6 Cam-driven, dual-head reciprocating-piston pump capable of delivering constant flow with relatively low pulsation. Flow rate is controlled by the cam rotation frequency. (Reprinted from Ref. 3 with permission.)

head pumps work on the same principle with the three pistons 120° out of phase, and, theoretically, flow pulsations are further minimized.

(ii) Pulse Dampers
Although pump pulsation is minimized in single-head pumps by making the refill time as short as possible, and in dual- or triple-head pumps by overlapping the piston movement, additional pulse damping is often required. Most reciprocating-piston pumps incorporate pulse dampers, or noise filters, between the pump and the injector. Pulse dampers are essentially long lengths of very narrow tubing folded back on themselves many times. These flow-through devices are operated by storing energy on their volume during the eluent delivery stroke and releasing the energy through a restrictor during the refilling stroke. By accumulating and discharging solvent in this way, minor flow fluctuations are smoothed out. The inclusion of a pulse damper, however, increases the system volume between the pump and the column and makes solvent changeover less convenient. For gradient elution, therefore, where accurate gradient profiles are important, a low-volume pulse damper is essential.

3.2.3 Optimization of Flow Reproducibility

To achieve the flow and pressure precision needed for the more demanding HPLC applications, a number of proprietary electropneumatic pressure and flow control loops have been created to ensure repeatable pump performance.

(i) Pressure Loops
Pressure loops are used to control the pump delivery rate, by the placement of a sensitive and highly responsive pressure transducer downstream of the pump heads. Once a normalized baseline is established, the sensing circuitry regulates motor speed to keep the pressure constant. A momentary decrease in pressure will cause the control loop to accelerate the motor in order to increase the pressure. Although the pressure loop is effective at maintaining a smooth pressure delivery, it is done at the expense of flow accuracy.

(ii) Flow Loops
Flow control loops are designed to maintain constant retention times by providing constant flow delivery from the pump, commonly by the use of an encoder wheel attached to the main motor drive. The encoder accurately measures the rotation of the motor extremely accurately to parts-of-a-degree of rotation, and it regulates the motor revolutions to as constant a value as possible. In a well-maintained pumping system, a constant motor speed will deliver a constant flow for a given solvent. In modern pumps a flow control loop is often combined with a pressure loop, to provide both

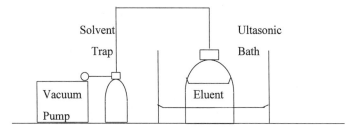

Figure 3.7 Basic external vacuum degassing station.

smooth and accurate solvent delivery. The combination of flow and pressure loops can provide accurate and essentially pulseless flow over a range of different solvent combinations and pressures.

3.3 Solvent Degassing

One of the more critical parameters, especially in gradient work, is the quality of the solvents used. In addition to being particulate and contaminant free, solvents must contain minimal levels of dissolved gases. Dissolved gases come out of solution when the eluents are pumped through the system, and the probability of gases bubbling out of solution increases when two or more liquids are mixed. The microbubbles frequently become attached to check valves and interfere with accurate flow. In addition, they can become lodged in detector cells, creating baseline instability. A number of ways to reduce or eliminate the problem are available.

3.3.1 External Vacuum Degassing

A simple and effective form of degassing is to hold a flask of mobile phase under a vacuum while agitating the contents in an ultrasonic bath (Fig. 3.7). The eluent is then transferred to the chromatograph for several hours of reliable operation, especially if the eluent is blanketed with an inert gas such as helium. This approach is particularly useful for clean solvents that readily absorb gases such as carbon dioxide from the atmosphere. The eluent should be degassed in the solvent storage container used for chromatography to minimize contact with the air.

3.3.2 Helium Sparging

Helium sparging is a simple technique: a helium tank is connected to the solvent reservoirs of the chromatograph, and helium is bubbled into

3.3 Solvent Degassing

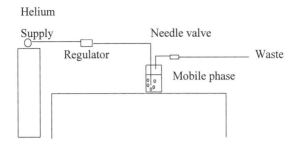

Figure 3.8 Helium sparge configuration.

the eluents. A specific, user-defined volume of helium is bubbled per minute for the duration of the analysis. If the chromatograph is not equipped with on-line sparging, the sparging can be performed off-line (Fig. 3.8). Because helium has very low solubility in most solvents, it displaces the more common gases from the mobile phase. This approach[4] has two limitations; not only is a large quantity of expensive helium gas required, but the potential also exists for changing the composition of the premixed solvents by selectively volatilizing the more volatile solvent.

3.3.3 On-line Degassing

Since the mid-1980s, on-line solvent vacuum degassing has gained considerable popularity, and such features are often included in pumping systems. With this approach a vacuum is drawn on semipermeable tubes through which the eluents run (Figure 3.9). The vacuum draws air from the flowing solvents and discards it to waste. When combined with inert gas blanketing, the resulting solvent can be used for the most demanding of gradient applications.

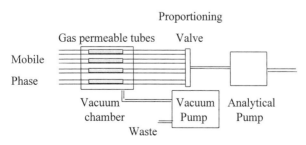

Figure 3.9 On-line degas schematic.

3.4 Isocratic versus Gradient Pumping Systems

3.4.1 Isocratic Pumping Systems

Isocratic elution is commonly used for the elution of analytes from the column. In isocratic elution, the mobile phase is kept constant throughout the analysis. The mobile phase can be a single solvent or a solution of two or more miscible solvents. The major requirements of isocratic pumps are accuracy and smoothness of flow. Because the pump delivers only one solvent system, simple, inexpensive pulse dampeners and rudimentary flow or pressure feedback control circuits can be used. The basic setup of an isocratic system is illustrated in Figure 3.10.

3.4.2 Gradient Pumping Systems

Gradient pumping systems are capable of delivering more than one solvent during an analysis. The blending of two solvents to form a gradient is referred to as a binary gradient, three solvents create a ternary gradient, and a quaternary gradient is formed by the use of four different solvents. The blending of the solvents can occur in one of two ways: high-pressure mixing or low-pressure mixing.

(i) High-Pressure Mixing

In high-pressure mixing the eluents are blended on the injector side, or "high-pressure" side, of the pump. Theoretically, several eluents can be mixed in this way; however, in practice only two eluents can be mixed conveniently, and two isocratic pumps are required, one for each solvent. The fluid lines of the two pumps are joined with a mixing device or T, as illustrated in Figure 3.11. The amount of each solvent in the gradient is controlled by the relative differences in the flow rates of the two pumps. High-pressure mixing is attractive for applications that demand very low delay volumes, such as in capillary LC or HPLC–mass spectrometry. These systems can be configured with delay volumes of only a few microliters. The most significant limitation of the high-pressure mixing configuration

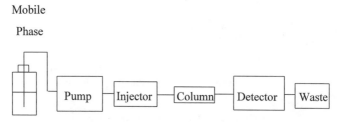

Figure 3.10 Schematic of an isocratic pumping system.

3.4 Isocratic versus Gradient Pumping Systems

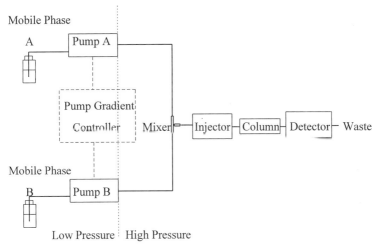

Figure 3.11   Typical high-pressure mixing configuration.

is that reciprocating pumps have poor precision at the extremes of the flow rate ranges, that is, in the early and late stages of the gradient. Another problem with high-pressure mixing is blending solvents whose compressibility or combined volumes are nonadditive. The classic example is a mixture of acetonitrile and water, where 100 ml of water added to 100 ml of acetonitrile yields only 170 ml of the mixture. Such a mixture created in a high-pressure mixing system will result in a reduction of flow rate owing to the compression of the solvents. Finally, with high-pressure mixing a complete pump is needed for each solvent, making the process costly for all but binary mixtures of solvents.

(ii) Low-Pressure Mixing

Low-pressure mixing has become a common design for modern gradient work. In this configuration the solvents are blended at atmospheric pressure, and a single high-pressure solvent delivery system is used to pump the mixture. By means of a small Teflon block with four proportioning valves, a controlled mixture of any combination of the four solvents is delivered to the "low pressure" intake side of the pump (Fig. 3.12). Because low-pressure mixing systems can provide delay volumes of less than 1 ml, they can be used with columns as small as 1 mm inner diameter (I.D.).

The criteria used to judge the performance of a low-pressure pump include compositional accuracy, ripple, flow accuracy, and pressure accuracy. Compositional accuracy is tested by following the American Society for Testing and Materials (ASTM) procedure E-19.09.07.[5] In this procedure, two bottles of eluent are used, one containing 100% methanol (eluent A) and the other containing methanol with a low concentration of acetone

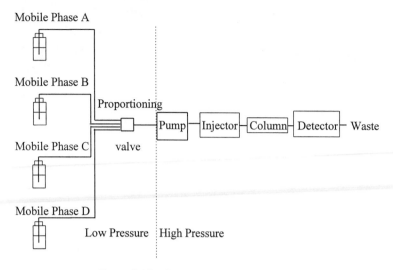

Figure 3.12 Low-pressure mixing system.

(eluent B) to act as a chromophore. Starting with 0% eluent B, equal step changes of increasing eluent B are programmed to 100% B as illustrated in Figure 3.13. The data are then inspected to determine how well the measured steps match those programmed. A reasonable delay volume and rounding of the staircase edges is inevitable to provide the needed mixing, but all the steps should be equal in height and at the concentration programmed; errors in the step heights indicate proportioning errors. The flat sections of each stair should be without "ripple," as ripple is indicative of inadequate mixing of the proportioned solvents. Excess ripple may be

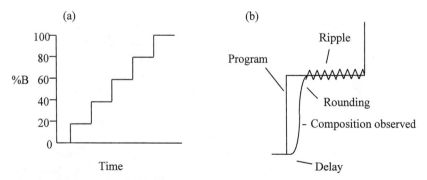

Figure 3.13 Representation of programmed step gradient showing (a) the ideal gradient and (b) typical gradient profile with rounding, ripple, and inconsistent steps highlighted.

reduced by the addition of mixers, but this will increase the delay volume of the system.

Flow accuracy across a range of backpressures and solvent compositions can be evaluated by collecting and weighing timed aliquots of solvent. Pressure readings are plotted on a strip-chart recorder from a transducer signal output point. The recorded values should be within the manufacturer's specifications and as consistent and close to the programmed values as possible.

3.5 Sample Introduction

The sample is introduced into the chromatograph via a sample injector. To minimize dispersion and the broadening of peaks, the sample must be injected as a sharp plug. Valve injectors are usually used in commercial instrumentation, because the sample is introduced with minimal interruption of flow. The injection devices are the basis for both manual and automatic injection because of their ease of use, reliability, and ease of automation.[6] A typical injection valve, the six-port valve, is illustrated in Figure 3.14.

The injection valve has two positions: the load position and the inject position. To introduce a sample, the valve is first switched to the load position, as shown in Figure 3.14a. In that position, mobile phase bypasses the sample loop and flows straight from the pump through the valve to the

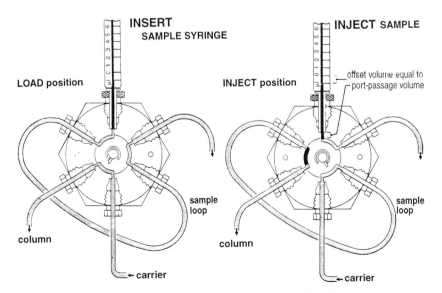

Figure 3.14 (a) Load and (b) Inject positions of a six-port valve. (Adapted from Ref. 6 with permission.)

column. At the same time, sample can be loaded into the sample loop through the needle port without interrupting the eluent flow. To introduce the sample into the chromatograph, the valve is switched to the inject position, as shown in Figure 3.14b, and the mobile phase is used to "backflush" the sample from the loop and into the column.

The sample loop size can be varied depending on the volume of sample. Typical sample loops range from 5 μl to 5 ml, and each must be manually removed before another can be put in place. The replacement of one loop size for another is one of the major disadvantages of the injector design. The accuracy of this type of injector varies with sample loop size, from about 5% for a 2-ml loop to as much as 30% for a 5-μl loop.

The need for unattended and precise sample injection for HPLC has lead to development of a wide variety of automated sample injection devices. Autosamplers function in essentially the same way as manual injectors, except that the sample is introduced automatically from a sample vial held in a carousel or an $X-Y$ grid (Fig. 3.15). The carousel format provides a reliable and rapid means of moving samples past an injection station, whereas the XY grid format allows a convenient "random access" configuration.

Sample is introduced into the sample loop in a variety of ways: the vial may be pressurized to force the sample out, or a syringe may be used to draw the sample out. The syringe is usually controlled by a stepping motor so that different sample volumes can be injected reproducibly by partially filling the sample loop. Once the sample loop has been loaded, the valve is electrically actuated. Some autosamplers also include other features such as sample heating or cooling, and the ability to perform standard additions, thereby improving the precision of the analysis.

Autosampler vials are available in different sizes, shapes, and materials. In terms of volume, vials from 300-μl ultracentrifuge tubes to 25-ml glass test tubes exist. For analytical work the standard autosampler vials are from 1 to 4 ml. Smaller vials, known as limited volume inserts, which fit inside the standard vials, are also available when the sample size is limited.

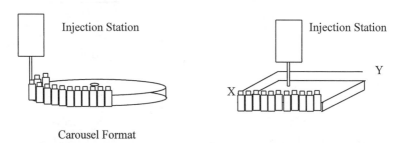

Figure 3.15 Typical carousel-based and XY-based injector configurations.

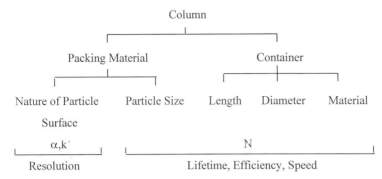

Figure 3.16 Components of a column and their contribution to chromatographic performance. (Adapted from Ref. 7 with permission.)

Most vials are made from clear borosilicate glass, but amber-colored vials are available for light-sensitive samples. Vials made from polymers such as polypropylene are also available if contaminants from the glass interfere with the analysis or if sample components adhere to the glass wall.

3.6 Columns

The column is the heart of the chromatograph, providing the means for separating a mixture into components. The selectivity, capacity, and efficiency of the column are all affected by the nature of the packing material or the materials of construction.[7] Figure 3.16 indicates how the components of the column contribute to chromatographic performance.

3.6.1 Stationary Phases

Most modern HPLC packings are microparticles of varying size, shape, and porosity. The surface of the particles can be modified by either physical or chemical means to afford access to any of the classic modes of chromatography. Silica gel is the main stationary phase for adsorption chromatography, although the use of other metal oxides,[8] such as alumina[9,10] and zirconia,[11] as well as carbon[12,13] and hydroxyapatite[14,15] have been reported. Polymeric adsorbents, such as polystyrene–divinylbenzene have also been used,[16] although these resins are usually used as supports for bonded phases.

Silica packings are popular because they can withstand the high pressures generated when 10- to 30-cm columns packed with 3- to 10-μm particles are used. Silica is abundant, inexpensive, and available in a variety of shapes, sizes, and degrees of porosity. In addition, functional groups can be readily bonded to the silanols, and the chemistry of the bonding reactions

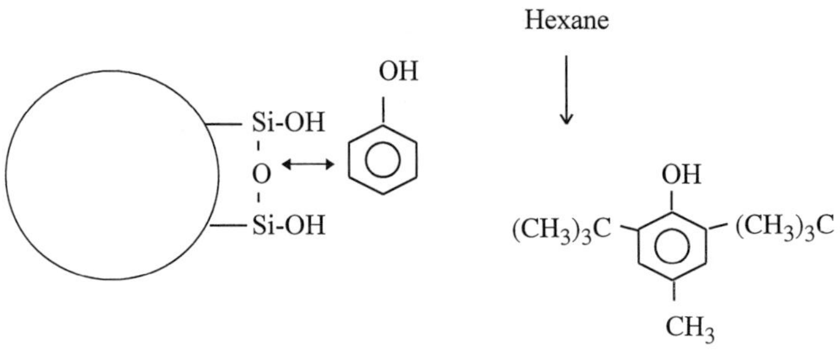

Figure 3.17 Normal-phase adsorption chromatography on silica.

are well understood. Figure 3.17 shows a schematic illustration of adsorption chromatography on silica. The main disadvantage of silica is its instability at high and low pH (i.e., above pH 8 or below pH 2), limiting the eluents that can be used.

Resin-based packings such as polystyrene–divinylbenzene and acrylic-based polymers are being used increasingly in HPLC columns. The packings are predominantly used in size-exclusion and ion-exchange chromatography; however, resin-based reversed-phase columns are also commercially available. Although resin-based columns can be used over a wide pH range (pH 1–13), they are more limited in terms of pressure than silica-based columns. Resin-based columns also tend to be more restricted in terms of organic modifier concentrations.[17] Figure 3.18 shows the structures of some commercially available resins.

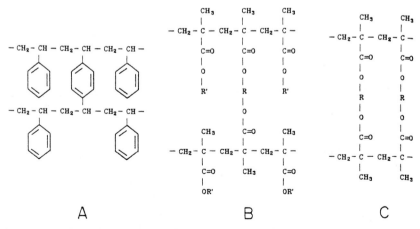

Figure 3.18 Commercially available resins: (A) polystyrene–divinylbenzene; (B), (C) polymethacrylate-based copolymers. (Reprinted from Ref. 16 with permission.)

3.6.2 Pore Size

Porous microparticles are the most common stationary phase particles used in modern HPLC. The role of pore size is a critical one, as the pores provide the surface with which the sample interacts. Particles with small pores exhibit a high surface area and therefore have greater retention. Large molecules like proteins, however, may be excluded from the small pores, and for those molecules a packing with a larger pore size is preferable. The difference between porous particles, pellicular particles, and porous microparticles is illustrated in Figure 3.19. Porous particles are seldom used owing to low efficiencies and are not discussed further.

Although pore sizes are usually represented by a single number, packing materials do not have discrete pore sizes. Instead, the packing is described as a statistical distribution of pore sizes and shapes. The distribution can be broad or narrow depending on the way in which the silica was made and the subsequent treatments it has undergone. For most modes of chromatography, a narrow distribution of pore sizes is desirable, but size-exclusion chromatography is the exception.

Pellicular materials consist of a solid spherical bead of relatively large

(a) Totally Porous Particle (20-40 μm)

Long pores filled with stagnant mobile phase

(b) Pellicular Particle (~20-40 μm)

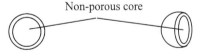

Non-porous core

Thin layer of adsorbent or stationary phase

(c) Totally Porous Microparticle (5-10 μm)

Short pores

Figure 3.19 Physical appearance of stationary-phase particles.

inner diameter with a thin outer surface of stationary phase. If the inner core is coated with a porous silica layer prior to being coated with the stationary phase, the particle is known as a superficially porous particle. Pellicular materials have emerged as the most common materials for ion-exchange chromatography. They give higher efficiencies (lower HETP) than porous particles of the same size but are restricted to small sample loadings because of the low active surface area.[17] Pellicular materials are more easily packed into the columns and therefore show advantages in terms of ease of packing, long-term stability, and lower cost.

Porous microparticles are fully porous materials that can be either irregular or spherical in shape. Owing to the many advantages of spherical microparticles, such as better stability at high pressures, larger sample volume capacity, and better detection sensitivity, irregular microparticles are seldom used. Porous microparticles are generally chosen for highest efficiency and speed, when peaks must be isolated for identification, for trace analysis, or for complex mixtures requiring a large peak capacity.

3.6.3 Particle Size

The particle size of the packing material determines the number of theoretical plates per unit length that can be generated (see Section 1.3.4). Small particle sizes result in high efficiencies; however, increased backpressure can occur as a result of the decreased column permeability. If the small particles are packed into shorter columns the speed of analysis will increase at the same efficiency.

3.6.4 Internal Diameter of Column

The internal diameter of a column will affect the sample load, the peak dilution, and the flow rate. The larger the inner diameter, the greater is the loading capacity and the higher is the flow rate. However, peak dilution increases with internal diameter, and therefore mass sensitivity decreases. Most analytical columns range from 2 to 5 mm in diameter. Narrow-bore (or smaller) columns with diameters of 2 mm or less are used for applications where high sensitivity is required, the amount of sample is limited, or solvent purchase and disposal costs are significant.

3.6.5 Column Length

Column length affects both the efficiency and the speed of the separation. Longer columns result in longer analysis times. However, the column efficiency tends to increase with length. In general, short columns are used for simple separations. Analytical columns can range from 30 to 300 mm in length.

3.6.6 Construction Materials of Column

The column must be constructed of materials that withstand the pressures used in HPLC and must also be chemically resistant to the mobile phase. Most columns are constructed from 316 stainless steel for rigidity and mechanical strength at high pressures. Glass, Teflon, and PEEK columns are also available for use with the more aggressive mobile phases, such as hydrochloric acid, or with solutes such as proteins that may adsorb to the stainless steel.[18] Polymeric columns are more common for ion-exchange packings, whereas glass is commonly used for protein separations. The requirements of an ideal HPLC column can be summarized into the ten general points listed in Table 3.1.

3.6.7 Column Care and Use

Because HPLC columns have a limited lifetime, care must be taken in their use. It is important to read the manufacturer's recommendations regarding eluents, flow rates, mobile phase pH, maximum pressure, organic modifier content, and storage procedures in order to prevent destruction of the column. Only HPLC-grade reagents, or better, should be used, and the eluent should be filtered prior to introduction into the chromatographic system. Guard columns should be used wherever possible to prevent con-

Table 3.1 Requirements for an Ideal HPLC Column

1. Particles should be spherical and available in particle diameters ranging from 3 to 10 μm
2. Particles should withstand typical pressures encountered during HPLC ((900–3000 psi (6.1–20.5 MPa) but ideally up to 4000 psi (27.2 MPa)) and should not swell or shrink with the nature of the eluent.
3. Particles should have a porosity in the range 50–70%, extending to 80% for size-exclusion chromatography.
4. Particles should contain no pores smaller than ~60 Å in diameter and should have a uniform pore size distribution.
5. Particles should be available with a range of mean pore diameters of 60–1000 Å.
6. The internal surface of the material should be homogeneous.
7. The internal surface should be capable of modification to provide a range of surface functionalities.
8. Packing materials should be chemically inert under all conditions of pH and eluent composition.
9. The physicochemical characteristics of the material should be reproducible from batch to batch and from manufacturer to manufacturer.
10. The material should be readily available and relatively inexpensive, and its chemical behavior should be well understood.

tamination of the column from the sample. Finally, the flow rate of the mobile phase should be changed in small increments in order to avoid sudden backpressure changes. Columns should not be backflushed unless indicated in the column manual, nor should they be stored in buffers, such as phosphate buffer, that promote microbial growth.

3.7 Detectors

The detector converts a change in the column effluent into an electrical signal that is recorded by the data system. Detectors are classified as selective or universal depending on the property measured. Selective (solute property) detectors, such as fluorescence detectors, measure a physical or chemical property that is characteristic of the solute(s) in the mixture; only those components which possess that characteristic will be detected. Universal (bulk property) detectors measure a physical property of the eluent. Thus, with refractive index (RI) detectors, for example, all the solutes which possess a refractive index different from that of the eluent will be detected. Selective detectors tend to be more sensitive than universal detectors, and they are much more widely used. Universal detectors are more commonly used in preparative chromatography, where a universal response is desired and sample size is large.

Detectors may also be classified according to whether they are destructive or nondestructive. A nondestructive detector is one in which the sample is unaltered by the detection process. The commonly used optical detectors, such as those that measure UV absorbance and RI, fall into the nondestructive category. Nondestructive detectors are often used in series to obtain extra qualitative information. Destructive detectors include electrochemical and mass spectrometric detectors in which at least part of the sample is altered by the detector itself.

3.7.1 Detector Properties

The ideal on-line detector has versatility, high sensitivity, the capacity for continuous monitoring of the column effluents, low noise level, wide linearity of response, stable baseline, insensitivity to flow rate and temperature changes, and response to all types of compounds. It is rugged, not too expensive, and is able to measure accurately a small peak volume without increasing its volume appreciably. The terms noise, sensitivity, and linearity are typically used in describing detector performance, as discussed below.

Noise is defined as variation of the output signal that cannot be attributed to the solute passing through the cell. It can result from many causes, including instrument electronics, temperature fluctuations, line voltage changes, and changes in flow rates. There are three forms of detector noise: short-term, long-term, and drift. Short-term noise widens the trace and appears as "fuzz" on the baseline, whereas with long-term noise the variation in the baseline appears as "valleys" and "peaks." The steady movement of the baseline either up or down is referred to as drift. Noise is discussed in more detail in Chapter 7.

Sensitivity is the ratio of detector response (peak height) to the sample concentration. The more sensitive the detector, the larger is the peak for a given solute concentration. More important than the size of the signal, however, is the signal-to-noise ratio (S/N), which measures the amount of signal visible above the baseline noise. If a particular detector produces a large signal, but is noisy, the sensitivity will be compromised. The sensitivity of an optical detector, such as an absorbance detector, may be enhanced by replacing the flow cell with another of longer path length. If this is done, the signal will be enhanced, but so will the noise. The detector with the greatest S/N ratio is the more sensitive detector. Detection limits are also discussed in detail in Chapter 7.

Accuracy is a measure of the closeness of a measurement to the true value. Precision is a measure of how reproducible the measurements are. For many detectors, the accuracy of a measurement is maintained by user calibration. For some detectors, however, such as photodiode array detectors, accuracy relies on internal calibration. The "linear dynamic range" of the detector is the maximum linear response, divided by the detector noise. The detector response is said to be linear if the difference in response for two concentrations of a given compound is proportional to the difference in concentration between the two samples. Most detectors become nonlinear as the sample concentration increases.

3.7.2 Absorbance Detectors

Absorbance detectors are nondestructive and respond only to substances that absorb radiation at the wavelength of the source light. They are classified as selective or solute property detectors because the mobile phase is chosen such that it exhibits little or no absorbance at the wavelength of interest. Detectors that measure only in the range 190–350 nm are ultraviolet (UV) absorbance detectors, whereas those that measure in the region 350–700 nm are visible (Vis) detectors. Detectors that span the range 190–700 nm are known as UV/Vis detectors. The UV or UV/Vis detectors are the most widely used, not only because of the relative insensitivity of the detector to temperature and gradient changes, but also because of the great number of compounds that absorb radiation in the UV range.

(i) Direct Absorbance Detection

With direct UV detection the eluent exhibits little or no absorbance at the wavelength to be monitored. When a solute is exposed to UV radiation, the radiation is absorbed by particular electronic configurations of the compound. Compounds with one or more double bonds absorb in the UV, as do compounds with unpaired electrons. The name given to a covalently unsaturated group, responsible for electronic absorbance, is "chromophore," and compounds containing these groups are said to be chromophoric. Table 3.2 lists absorbance data for isolated or nonconjugated chromophores.[19]

The function of the detector is to monitor the light that passes through the eluent. When a UV-absorbing compound, dissolved in the eluent, passes through the detector, it absorbs some of the light, thereby preventing it from reaching the light sensor. The decrease in light results in an electrical output that is fed from the detector to the data system. Solute concentration and the intensity of light transmitted through the flow cell are related according to the Lambert–Beer law (Beer's law). Because absorbance is greater when a cell with a longer path length is used, flow cells with longer path lengths should be used to increase the sensitivity of the detector. In addition, the greater the molar absorptivity of the solute and/or the greater the solute concentration, the greater will be the signal.

There are several designs available for UV detectors. The earliest type was the fixed-wavelength detector, which, as the name suggests, is capable of monitoring one wavelength only. A dual-channel, fixed-wavelength detector has also been designed; however, despite the advantage of having two wavelengths to monitor, the noise of the detectors is considerably worse than in single-channel, fixed-wavelength detectors. In a single-channel, fixed-wavelength detector, monochromatic light of a specific wavelength passes through the flow cell to the detector. Very little stray light reaches the detector, thereby minimizing detector noise. In addition, most of the light originating from the lamp reaches the detector or is absorbed by a chromophoric group, with little light being lost to the surroundings. This detector is the most sensitive of UV detectors, as well as the least expensive. However, it has the disadvantage that only one wavelength can be monitored.

Some of the more common light sources used in fixed-wavelength detectors are listed in Table 3.3. Different wavelengths are selected by use of an appropriate filter or grating orientation. The most common line source is the low-pressure mercury lamp, which emits 90% of its light at 254 nm. Many compounds of biological importance, such as aromatic amino acids, proteins, enzymes, and nucleic acid constituents, absorb strongly at 254 nm, making this detector extremely effective.

Variable-wavelength and photodiode array detectors have the advantage that more than one wavelength can be monitored during a single run. The detectors use a light source with a continuous emission spectrum, such as a deuterium lamp, and a continuously adjustable narrow-band filter, known as a monochromator. The variable-wavelength detector can be programmed to monitor different wavelengths at different times during the

Table 3.2 Absorbance Data for Isolated Chromophores[a]

Chromophoric group	System	Example	λ_{max}(nm)	ε_{max}	Transition	Solvent
Ethylenic	RCH=CHR	Ethylene	165	15,000	$\pi \to \pi^*$	Vapor
			193	10,000	$\pi \to \pi^*$	
Acetylenic	R—C≡C—R	Acetylene	173	6,000	$\pi \to \pi^*$	Vapor
Carbonyl	RR'C=O	Acetone	188	900	$\pi \to \pi^*$	n-Hexane
			279	15	$n \to \pi^*$	
Carbonyl	RHC=O	Acetaldehyde	290	16	$n \to \pi^*$	Heptane
Carboxyl	RCOOH	Acetic acid	204	60	$n \to \pi^*$	Water
Amido	RCONH$_2$	Acetamide	<208	—	$n \to \pi^*$	—
Azomethine	>C=N—	Acetoxime	190	5,000	$n \to \sigma^*$	Water
Nitrile	—C≡N	Acetonitrile	<160	—	$\pi \to \pi^*$	—
Azo	—N=N—	Azomethane	347	4.5	$n \to \pi^*$	Dioxane
Nitroso	—N=O	Nitrosobutane	300	100	$n \to \pi^*$	Ether
			665	20		
Nitrate	—ONO$_2$	Ethyl nitrate	270	12	$n \to \pi^*$	Dioxane
Nitro	—N$\begin{smallmatrix}O\\O\end{smallmatrix}$	Nitromethane	271	18.6	$n \to \pi^*$	Alcohol
Nitrite	—ONO	Amyl nitrite	218.5	1,120	$\pi \to \pi^*$	Petroleum ether
			346.5[b]		$n \to \pi^*$	
Sulfoxide	S=O	Cyclohexyl methyl sulfoxide	210	1,500		Alcohol
Sulfone	>S$\begin{smallmatrix}O\\O\end{smallmatrix}$	Dimethyl sulfone	<180	—		—

[a] Reprinted from Ref. 19 with permission.
[b] Most intense peak of fine structure group.

Table 3.3 Common UV-Detection Wavelengths

Lamp	Wavelength (nm)
Mercury	254, 280, 365
Zinc	214, 308
Cadmium	229, 326

chromatographic process, to ensure that each peak is monitored at its maximum absorbance wavelength. With a photodiode array detector or diode array detector (PDA, DAD), either a single wavelength from a selected photodiode or the entire spectrum from the whole photodiode array can be monitored.

With variable-wavelength detectors, more light is lost to the surroundings than with single-wavelength detectors; thus, a single-wavelength UV detector is more sensitive than the other detectors. In addition, because more than a single wavelength can be monitored, the price of variable-wavelength detectors is higher. However, most current, commercially available variable-wavelength and photodiode array detectors have excellent sensitivity. Since the late 1980s, variable-wavelength detectors and photodiode array detectors are more commonly used than single-wavelength detectors, as so much more information is obtainable in a single run. Nevertheless, fixed-wavelength detectors are used still.[20] Typical eluents and absorbances at specified wavelengths are listed in Table 3.4.[21]

(ii) Indirect Absorbance Detection

If a solute of interest does not contain a chromophore, it may be detected by indirect UV detection. Indirect detection is a technically simple and sensitive method for the detection of compounds with little inherent detector response. Indirect UV detection is a nondestructive technique, in which the analyte is characterized in native form. Indirect detection is a universal detection mode, with few requirements as to the exact nature of the analyte. The properties of indirect detection have been reviewed by Yeung.[22] Indirect detection is particularly attractive for the analysis of biological compounds. Optical systems are the same for direct or indirect detection; the only difference is that, in indirect detection, the mobile phase, rather than the analyte, contains a UV chromophore.

The application of indirect methods in biomedical analysis has been reviewed by Schill and Arvidsson.[23] Reversed-phase HPLC is the main field of application for indirect detection, and both charged and uncharged species can be visualized, although sensitivity is better for ionic solutes. With indirect UV detection for reversed-phase HPLC, the eluent contains a chromophoric group (probe), and peaks are obtained for injected solutes as well as for the mobile phase additives (system peaks). For solutes that have the opposite charge from the probe, as shown in Figure 3.20a,[24,25] negative peaks are produced if the peaks are eluted before the system peak, and positive peaks result if the solutes are eluted after the system peak. If the solutes are uncharged, or have the same charge as the probe, the reverse is seen. Positive peaks are produced if the solutes are eluted before the system peak, but negative peaks result if they are eluted after the system peak, as illustrated in Figure 3.20b. These rules apply without exception if the mobile phase contains only a hydrophobic probe dissolved in an aqueous buffer with highly hydrophilic components.[23]

Typical probes for the analysis of ionic solutes include 3-hydroxy-L-tyrosine (DOPA)[24] and naphthalene-2-sulfonate,[26] whereas those for use with uncharged solutes include nicotinamide,[27] theophylline,[28] and anthracene.[29] Indirect detection is nonspecific and less suitable for the analysis of complex or impure samples, because unpurified biological samples, such as urine, contain a large number of hydrophilic solutes that will give problems such as extra system peaks. However, analyses of pharmaceutical products and quantification of impurities in substances are typical of applications.[23]

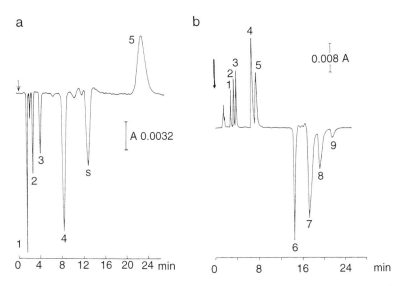

Figure 3.20. Analysis of carboxylic acids and alcohols by reversed phase HPLC, with indirect UV detection. (a) Carboxylic acids. Chromatography conditions: mobile phase, 3×10^{-4} M 1-phenethyl-2-picolinium in acetate buffer (pH 4.6); column, μ-Bondapak phenyl; detection, indirect UV absorbance at 254 nm. Peaks: 1, acetic acid; 2, propionic acid; 3, butyric acid; 4, valeric acid; 5, caproic acid; S, system peak. (b) Aliphatic alcohols. Chromatography conditions: mobile phase, 4×10^{-4} M nicotinamide in water; column, Ultrasphere ODS; detection, indirect UV absorbance at 268 nm. Peaks: 1, methanol; 2, propylene glycol; 3, ethanol; 4, 2-propanol; 5, 1-propanol; 6, system peak; 7, 2-butanol; 8, 2-methyl-1-propanol; 9, 1-butanol. (Redrawn from Refs. 23 and 24 with permission.)

Table 3.4 Mobile-Phase Spectral Data[a]

	Absorbance (AU) at specified wavelength (nm)								
	200	205	210	215	220	230	240	250	260	280
Solvents										
Acetonitrile	0.5	0.03	0.02	0.01	0.01	<0.01				
Methanol	2.06	1.00	0.53	0.37	0.24	0.11	0.05	0.02	<0.01	
Degassed	1.91	0.76	0.35	0.21	0.15	0.06	0.02	<0.01		
2-Propanol	1.80	0.68	0.34	0.24	0.19	0.08	0.04	0.03	0.02	0.02
Tetrahydrofuran										
Fresh	2.44	2.57	2.31	1.80	1.54	0.94	0.42	0.21	0.09	0.05
Old	>2.5	>2.5	>2.5	>2.5	>2.5	>2.5	>2.5	>2.5	2.5	1.45
Acids and bases										
Acetic acid, 1%	2.61	2.63	2.61	2.43	2.17	0.87	0.14	0.01	<0.01	
Hydrochloric acid, 6 mM (0.02%)	0.11	0.02	<0.01							
Phosphoric acid, 0.1%	<0.01									
Trifluoroacetic acid										
0.1% in water	1.20	0.78	0.54	0.34	0.20	0.06	0.02	<0.01		
0.1% in acetonitrile	0.29	0.33	0.37	0.38	0.37	0.25	0.12	0.04	0.01	<0.01
Ammonium phosphate, dibasic, 50 mM	1.85	0.67	0.15	0.02	<0.01					
Triethylamine, 1%	2.33	2.42	2.50	2.45	2.37	1.96	0.50	0.12	0.04	<0.01
Buffers and salt										
Ammonium acetate, 10 mM	1.88	0.94	0.53	0.29	0.15	0.02	<0.01			
Ammonium bicarbonate, 10 mM	0.41	0.10	0.01	<0.01						
EDTA (ethylenediaminetetraacetic acid), disodium, 1 mM	0.11	0.07	0.06	0.04	0.03	0.03	0.02	0.02	0.02	0.02

Component								
HEPES [N-(2-hydroxyethyl)piperazine-N-2-ethanesulfonic acid], 10 mM (pH 7.6)	2.45	0.50	2.37	2.08	1.50	0.29	0.03	<0.01
MES [2-(N-morpholino)ethanesulfonic acid], 10 mM (pH 6.0)	2.42	2.38	1.89	0.90	0.45	0.06	<0.01	
Potassium phosphate								
Monobasic, 10 mM	0.03	<0.01						
Dibasic, 10 mM	0.53	0.16						
Sodium acetate, 10 mM	1.85	0.96	0.05	0.01	<0.01			
Sodium chloride, 1 M	2.00	1.67	0.52	0.30	0.15	0.03	<0.01	
Sodium citrate, 10 mM	2.48	2.84	0.40	0.10	<0.01	0.54	0.12	0.03
Sodium formate, 10 mM	1.00	0.73	2.31	2.02	1.49	0.03	<0.01	0.02
Sodium phosphate, 100 mM (pH 6.8)	1.99	0.75	0.53	0.33	0.20	0.01	0.01	0.01
Tris–hydrochloric acid, 20 mM			0.19	0.06	0.02			<0.01
pH 7.0	1.40	0.77	0.28	0.10	0.04	<0.01		
pH 8.0	1.80	1.90	1.11	0.43	0.13	<0.01		
Detergents								
Brij 35 (23 lauryl ether), 1%	0.06	0.03	0.02	0.02	0.02	0.01	<0.01	0.02
CHAPS {3[(3-cholamidopropyl)dimethyl-ammonio]-1-propanesulfonate}, 0.1%	2.40	2.32	1.48	0.80	0.40	0.08	0.04	0.01
SDS (sodium dodecyl sulfate), 0.1%	0.02	0.01	<0.01					
Triton X-100 (octoxynol), 0.1%	2.48	2.50	2.43	2.42	2.37	2.37	0.50	0.25	1.42
Tween 20 (polyoxyethylenesorbitan monolaurate), 0.1%	0.21	0.14	0.11	0.10	0.09	0.06	0.05	0.04	0.03

[a] Reprinted from Ref. 21 with permission.

3.7.3 Fluorescence Detectors

Fluorescence is a specific type of luminescence. Fluorescent molecules absorb light at one wavelength, and then reemit it, essentially instantaneously, at a longer wavelength (lower energy). Fluorescence is expected in molecules that are aromatic or contain multiple conjugated double bonds with a high degree of resonance stability. In either case, the compound will possess delocalized π electrons that can be placed in low-lying excited singlet states. Polycyclic aromatic systems, which possess a great number of delocalized π electrons, are usually more fluorescent than benzene and benzene derivatives. Electron-donating groups present on the ring, such as $-NH_2$, $-OH$, $-F$, $-OCH_3$, and $-N(CH_3)_2$, tend to enhance fluorescence, whereas electron-withdrawing groups, such as $-Cl$, $-Br$, $-I$, $-NO_2$, and $-CO_2H$, tend to decrease or quench fluorescence. Thus, aniline fluoresces but nitrobenzene does not. Molecular rigidity is also important for fluorescence; in general, aromatic compounds that are the most planar, rigid, and sterically uncrowded are the most fluorescent. Occasionally, molecules with nonbonding electrons, such as amine groups, will fluoresce, but usually a delocalized π system must be part of the molecule.

Because fluorescence detection is an optical technique, it is also subject to Beer's law. For dilute solutions, where $\varepsilon c l < 0.01$,

$$I_f = 2.303 \Phi_f I_0 \varepsilon c l \tag{3.1}$$

where I_f is the intensity of fluorescence, Φ_f is the quantum efficiency of fluorescence (the number of photons emitted divided by the number absorbed), I_0 is the incident intensity of light, ε is the molar absorptivity of the absorbing species, c is the concentration of the compound (moles per liter), and l is the cell path length. Equation (3.1) indicates a linear relationship between the intensity of fluorescence and concentration of the solute, provided that $\varepsilon c l$ is small. It also illustrates that the dependence of fluorescence intensity on path length is much less critical than it is in absorption detection. Of particular interest, however, is the linear dependence of fluorescence on the excitation power. This means that sensitivity can be increased by working at a high excitation power to give a large signal-to-noise ratio. The use of lasers as excitation sources instead of conventional lamps provides a number of advantages. The most obvious is the increase in sensitivity that is seen with laser-induced fluorescence. In addition, because of the coherent and monochromatic nature of the laser beam, little light is lost to the surroundings.

In fluorescence detection the sample absorbs light and then reemits it in all directions. In principle, therefore, the fluorescent light can be observed from any angle. Although other geometries are encountered, a 90° angle is used most frequently in commercial instruments because the arrangement minimizes any incident light that could provide a background signal to the fluorescence sensor. Fluorometers that use filters to select the excitation and emission wavelengths are called filter fluorometers, whereas those that use a monochromator are known as spectrofluorometers. Filter fluorometers are the less expensive of the two designs, and they frequently pass light in a wider band than do monochromators. Spectrofluorometers, however, have the advantage that both the excitation and the emission wavelengths can be changed.

The mobile phase plays an important part in the fluorescence of a molecule. Unless chosen with care, the mobile phase can quench the fluorescence of the molecule of interest. Most of the non-halogen-containing solvents used in HPLC can be used with fluorescence detection. However, dissolved oxygen or other impurities in the eluent can cause quenching. The solvent polarity and the pH of the mobile phase can also affect the fluorescent process if they influence the charge status of the chromophore. For example, aniline fluoresces at pH 7 and at pH 12, but at pH 2, where it is cationic, it does not fluoresce. Table 3.5 shows wavelength selections for some common LC–fluorescence applications.[30]

Table 3.5 Wavelength Selections for Common LC–Fluorescence Applications[a]

Application	Mobile phase	Wavelength (λ, nm)	
		Excitation	Emission
Aflatoxin G_1, G_2	Hexane/tetrahydrofuran	363	425–440
Benzo[a]pyrene	Water/acetonitrile	360	460
LSD	Aqueous ammonium carbonate/acetonitrile	330	460
Fluorescamine derivatives	Aqueous buffers	400	500
O-Phthalaldehyde derivatives	Aqueous buffers	360	460
Dansyl amino acids	Aqueous sodium acetate/ acetic acid buffer/ acetonitrile	360	490
Dansyl polyamines	Water/acetonitrile	360	490
Dansyl cannabinoids	Hexane/dichloro- methane/methanol	360	490
Indoles	Water	280	335
Quinolines	Water	350	500
Quinine	Water plus sulfuric acid	360	490
Estrogen steroids	—	285	325

[a] Reprinted from Ref. 30 with permission.

(i) Derivatization

Chemical derivatization of an analyte is performed to improve the selectivity and sensitivity of that analyte. There are two approaches to the use of derivatizing agents in HPLC. Precolumn derivatization, which is performed before the analytes are separated, is most commonly used for small molecules such as amino acids, whereas postcolumn derivatization, which is performed after the separation but before the analytes reach the detector, is often used for larger molecules such as peptides.

There are several points to consider when choosing a derivatizing scheme, such as the stability of the derivative to hydrolysis, solvolysis, and thermal decomposition. In addition, if precolumn derivatization is chosen, the derivatization process will alter the chromatographic properties of the analytes, which may result in the need to adopt a different chromatographic mode. Finally, if postcolumn derivatization is chosen, the effects of band broadening and sample loss caused by adsorption and dilution effects should not be overlooked.

Derivatization is a technique that is most commonly performed prior to UV absorption or fluorescence detection. It is not restricted to these detection modes, however, and postcolumn electrochemical[31] and postcolumn chemiluminescence[32] detection have also been reported. Table 3.6 provides a list of some of the more common derivatizing agents and the compounds with which they react.[33] Derivatization has also been used to aid in the detection of compounds such as amino acids,[34] amines,[35] saccharides,[36] thiols,[37] carboxylic acids,[38] steroids,[39] alcohols,[40] fatty acids,[41] and several inorganic species.[42]

A. Precolumn Derivatization Precolumn derivatization tends to be the more sensitive of the two methods, but it requires additional steps prior to the chromatographic separation of the solutes in the sample. However, standard chromatography equipment can be used without modification. Because the precolumn derivatization requires additional time on the part of the chemist, precolumn derivatization is best suited for situations where derivatization is not the standard protocol.

In precolumn derivatization, the derivatization process alters the chemical nature of the analytes. Therefore, it may be necessary to develop new chromatographic methods. However, the separation can be optimized for the particular analytes, and any excess reagent can be removed so that it does not interfere with detection. The selection of precolumn derivatization reagents is therefore less restricted than is the choice of postcolumn derivatization reagents, and rapid kinetics are not particularly important. The stability of the derivative is important, however, as is the percent derivatization, which should be as near to 100% as possible. It is also important that the reaction yield only one derivative per analyte, so that coelutions of extra peaks does not occur, and so solute identification and quantitation are accurate.

3.7 Detectors

Table 3.6 Fluorescent Derivatizing Reagents[a]

Reagent	Reacts with	Excitation λ (nm)[b]	Emission λ (nm)[b]
Fluorescamine	Primary amines	390	475
o-Phthalaldehyde	Primary amines	338	425
Naphthalenedialdehyde	Primary amines	442	490
CBQCA[c]	Primary amines	442	550
Fluorescein isothiocyanate (FITC)	Primary amines	488	525
AQC[d]	Primary and secondary amines	250	395
NBD chloride	Primary and secondary amines	420	540
Dansyl chloride[e]	Primary and secondary amines, phenols	360	520
Dansyl hydrazine	Aldehydes and ketones	340	525
Bromomethylcoumarin	Carboxylic acids	325	430
Phenylglyoxal	Guanine, guanine nucleosides and nucleotides	365	515
Thiazole orange	DNA intercalator	488	520

[a] Adapted from Ref. 33 with permission.
[b] Emission and excitation wavelengths are solute- and solvent dependent. A number of the excitation wavelengths correspond to the laser lines (325, 442, and 488 nm) and do not correspond to the actual excitation maxima. In addition, the excitation wavelength is frequently selected to avoid the Raman band and may not correspond to the actual emission maximum.
[c] 3-(4-Carboxybenzoyl)-2-quinolinecarboxaldehyde, available from Molecular Probes.
[d] 6-Aminoquinolyl-N-hydroxysuccinimidyl carbamate, available from Waters Chromatography.
[e] Dansyl chloride has poor quantum yield in water.

Precolumn derivatization has been applied to the analysis of many compounds, including amino acids (Table 3.6), drugs using cationic micelles,[43] and metals using Eriochrome B.[42]

B. Postcolumn Derivatization Three types of reactors for postcolumn derivatization are used, depending on reaction kinetics. Straight, coiled, and knitted open-tubular reactors are used for fast reactions, whereas packed-bed reactors are used for intermediate kinetics. Segmented-stream reactors are used for slow reactions. The simplest reactors are the open-tubular reactors; a T connector is the most common. Pickering[44] has described the performance requirements for instrumental components of HPLC postcolumn systems.

Postcolumn derivatization usually requires a fast reaction time. To keep the reagent blank interference to a minimum, a reagent must be used that does not share the detection properties of the derivative. Because the analytes are separated in native form, published methods can be used for the separation and new methods need not be developed. A major advantage of postcolumn derivatization is that it is not important that the reaction produce a single product, as multiple products will be eluted together and detected as a single compound. Postcolumn derivatization is the best choice for routine analyses, since it is amenable to automation; however, it is generally less sensitive than precolumn derivatization, it requires more complex hardware, and the analysis times are usually longer.

Postcolumn derivatization has been applied to the detection of numerous compounds including arginine compounds using ninhydrin,[45] guanine and guanine nucleosides and nucleotides using phenylglyoxal reagent,[46] cysteine-containing compounds using fluorescein isothiocyanate,[47] and proteins using indocyanine green dye.[48]

3.7.4 Electrochemical Detectors

Electrochemical detectors measure the current resulting from the application of a potential (voltage) across electrodes in a flow cell. They respond to substances that are either oxidizable or reducible and may be used for the detection of compounds such as catecholamines, carboxylic acids, sulfonic acids, phosphonic acids, alcohols, glycols, aldehydes, carbohydrates, amines, and many other sulfur-containing species and inorganic anions and cations. Potentiometric, amperometric, and conductivity detectors are all classified as electrochemical detectors.

The conductivity detection mode measures the change in conductance in the solution between two electrodes caused by the introduction or removal of charged species. In the amperometric detection mode, on the other hand, compounds undergo oxidation or reduction reactions through the loss or gain, respectively, of electrons at the electrode surface. The electrical current arising from the electrons passed to or from the electrode is recorded and is proportional to the concentration of the analyte present. Figure 3.21 illustrates the difference between amperometry and conductivity.[49]

(i) Conductivity Detectors

Conductivity detectors are universal and nondestructive, and they can be used in the direct[50] or indirect[51] modes. The separation of a group of anions and a group of cations by indirect conductivity detection is shown in Figure 3.22. The basis of conductivity detection is the forcing of ions in solution to move toward the electrode of opposite charge on application of a potential. The conductance, which is the reciprocal of the electrical resistance, depends directly on the number of charged species in the solution. Conductivity detection is the detection mode of choice for organic or inorganic species that are charged when they enter the detector cell.

3.7 Detectors

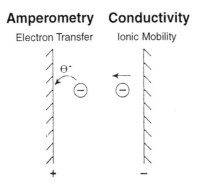

Figure 3.21 Properties measured during electrochemical detection. Amperometry measures the current or charge transferred between neutral or ionic analytes and the electrode. Conductivity measures the mobility of ions in an electric field. (Reprinted from Ref. 49 with permission.)

Because the species detected by conductivity detection are ionic, ion-exchange and ion-pair chromatography are the separation modes most often used. Both methods require mobile phases containing strong electrolytes; however, the detector must detect the ionic solutes without being over-

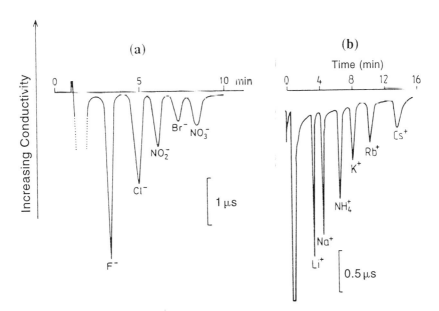

Figure 3.22 Indirect conductivity detection of (a) anions, using a TSK-GEL 620 SA column with 2 mM KOH as the eluent, and (b) cations, using an IC-PAK C column with 2 mM HNO$_3$ as the eluent. (Reprinted from Ref. 51 with permission.)

whelmed by the ions in the mobile phase. A solution to the problem has been to suppress the conductivity of the mobile phase.[52] Suppression is achieved in anion-exchange chromatography by exchanging the eluent cation with hydronium ions as the effluent leaves the column but before it reaches the detector. The hydronium ions combine with the anionic component of the mobile phase, effectively suppressing its conductivity. In cation-exchange chromatography, the anion in the mobile phase is exchanged for hydroxide ions, which neutralize the conductivity of the cationic component of the mobile phase.

Mobile phases useful for suppressed conductivity detection of anions include sodium hydroxide, potassium hydroxide, and the sodium and potassium salts of weak acids such as boric acid. In nonsuppressed conductivity detection, the ionic components of the mobile phase are chosen so that their conductivities are as different from the conductivity of the analyte as possible. Large ions with poor mobility are often chosen, and borate-gluconate is popular. For cations, dilute solutions of a strong acid are often used for nonsuppressed conductivity detection. For more information on the application of electrochemical detection to inorganic analysis, see *Ion Chromatography: Principles and Applications* by Haddad and Jackson,[17] which provides a comprehensive listing of the sample types, analytes, sample pretreatments, columns, and mobile phases that have been used with electrochemical detection.

Because the sensitivity of the detector decreases with decreasing analyte ionization, the pH of the mobile phase should be chosen to maximize solute dissociation. For example, anions with pK_a values above 7 are not detectable by conductivity detection. However, conductivity detection is often the preferred method for organic acids with carboxylate, sulfonate, or phosphonate functional groups, since the pK_a values are below 5. For cations, most aliphatic amines have pK_a values around 10 and are readily detected by conductivity detection. The pK_a values of aromatic amines, however, are in the range 2 to 7, which is too low to be detected by suppressed conductivity. Sensitivity by nonsuppressed conductivity is also poor, so these amines are monitored by UV absorption or pulsed amperometric detection.

(ii) Amperometric Detectors

Amperometric detection is used to measure the current or charge resulting from oxidation or reduction of analyte molecules at the surface of the working electrode. It is a selective and destructive technique. Oxidation or reduction of analyte molecules is accomplished by applying a potential between the working and reference electrodes in the amperometry cell. The working electrode is typically glassy carbon, or a precious metal electrode such as gold, silver, or platinum. The choice of working electrode is determined by potential limits in the mobile phase, the kinetics of the transfer reaction, and involvement of the electrode in the electrochemical reaction.

The negative potential limit is the potential at which the mobile phase is reduced, and the positive potential limit is the potential at which it is oxidized. The potential used to detect the analyte must be between the two limits. The noise will increase as the applied potential for the reaction approaches the potential limit, owing to the increased background current. Potential limits are strongly affected by the pH of the mobile phase, so that the negative potential limits are more negative in base and more positive in acid. Conversely, positive potential limits are more positive in acid and more negative in base. Glassy carbon electrodes provide both the largest positive and the largest negative potentials. Although large positive potentials are obtained with platinum electrodes, they have poor negative potential limits and are generally not used for reductions.

When the purpose of the working electrode is to act as an inert electron sink, as in the detection of catecholamines, carbon is the preferred electrode material. On occasions when the electrode plays a direct role in the reaction, the precious metals are chosen. For example, silver can be oxidized to silver cyanide in the presence of cyanide ions. A major consideration when choosing an electrode material is its ability to maintain an active surface. Electrodes will develop a layer of surface oxide at positive applied potentials. The oxide layer will inhibit the oxidation of the analyte, and the response will decrease with repeated injections. The active surface can be renewed by polishing the electrode. Since glassy carbon electrodes are more resistant to poisoning by oxide formation, they are the electrode of choice for direct current amperometry.

Direct current (DC) amperometry is used for the analysis of catecholamines, phenols, and anilines, which are easy to oxidize. A single potential is applied, and the current is measured. The current resulting from the oxidation or reduction of analyte molecules is dependent on many factors, including the concentration of the analyte, temperature, the surface area of the working electrode, and the linear velocity of the flowing stream over the surface of the working electrode.

Pulsed amperometric detection is used for the analysis of carbohydrates and other nonchromophoric molecules such as alcohols, aldehydes, and amines. In the DC mode the products of the oxidation reaction of those compounds poison the surface of the working electrode, and further analyte oxidation is inhibited. This results in peaks that decrease rapidly in height. To maintain a stable and active electrode surface, alternatively positive and negative potentials are repeatedly pulsed.

3.7.5 Mass Spectrometric Detectors

A mass spectrometer is a universal detector that is used for the analysis of compounds with molecular weights up to and in excess of 100,000 amu. It can be used to confirm the identity of a compound and will often provide sufficient data to determine the structure of an unknown. Although the mass spectrometer is not the most sensitive of HPLC detectors, it is only an order of magnitude less sensitive than laser-induced fluorescence detection. A mass spectrometer bombards the substance under investigation with an electron beam and quantitatively records the result as a spectrum of positive fragments (the mass spectrum). The mass spectrometer detector is therefore a universal, destructive detector.

The quadrupole mass spectrometer is the detector of choice for interfacing with chromatographs, and soft ionization methods, such as those listed in Table 3.7, are the most popular.[53] Owing to their ability to generate gaseous ions from large, nonvolatile, and labile molecules, soft ionization methods are the best choice for the analysis of biopolymers. Electrospray mass spectrometric detectors are of particular interest, because multiply charged species (as opposed to singly charged species for other techniques) can form if the analyte contains more than one site for cation attachment. The phenomenon allows the molecular weight range of biopolymers amenable to the technique to be vastly increased. For compounds with molecular weights below 1000, however, ion spray is more popular than electrospray. The advantages and disadvantages of several ionization methods are listed in Table 3.7.

The choice of interface is dependent on both the particular analysis and the instrumentation available. Some interfaces require the use of very low flow rates and therefore necessitate the use of either microbore or capillary LC equipment, or a sample splitter if standard-bore equipment is used. Thermospray ionization is the most frequently quoted interface, owing to compatibility with standard-bore instruments. However, the upper molecular weight limit for thermospray ionization is low, and the electrospray interface is becoming popular. The maximum flow rates for different interfaces are listed in Table 3.8.[54]

Mass spectrometry (MS) detection for HPLC may be performed using a single-quadrupole or a triple-quadrupole mass spectrometer. In LC–MS, only one quadrupole is used, and the data are employed for identification or qualitative information. Full-scan spectra may be obtained, where mass-to-charge ratios (m/z) are sequentially scanned to produce a spectrum over the desired m/z range. In LC–MS–MS, a triple quadrupole is used to obtain more information than is possible with LC–MS. In the fragment ion mode, a spectrum of structurally significant fragment ions is obtained, resulting in a "fingerprint" for the preselected precursor ion. Precursor-ion scanning and neutral-ion scanning modes are frequently used in the screening of unknown metabolites and drugs belonging to the same class, where the analyst can observe different molecular ions generated from common specified fragment ions or compounds showing the loss of a common fragment ion.[55]

High-performance liquid chromatography with mass spectrometry detection has been applied to the analysis of compounds such as azo dyes in wastewater, trace levels of pesticides in water, and picogram levels of

Table 3.7 Analysis of Peptides and Proteins by Mass Spectrometry[a]

Ionization method	Upper mass limit (amu) for MW information	Advantages	Disadvantages
Fast atom bombardment	20,000	Ability to analyze mixtures; MW and limited sequence information for pure small peptides; adaptable for LC/MS (NP- and RP-HPLC); widely available	Matrix interference below m/z 500; widely variable response; peptide-induced suppression effects; limited to flow rates ≤ 10 $\mu l/min$
Electrospray ionization	$\geq 150,000$	Ability to analyze mixtures; limited sequence information for pure small peptides; adapts easily for LC/MS (NP- and RP-HPLC); most suitable for quadrupoles; multiple charging increases upper mass limit; well suited to polar or ionic compounds	Multiple charging may complicate interpretation of data; glycoproteins may not yield useful information; limited to flow rates ≤ 10 $\mu l/min$
Matrix-assisted laser desorption	>230,000	Highest achieved mass range for proteins and glycoproteins; relatively insensitive to salts; ability to analyze mixtures; simple to operate	Very low resolution limits; ability to detect structural variants; not presently adapted for LC/MS; limited structural information
Plasma desorption	<50,000	Simple to operate; ability to analyze mixtures; limited sequence information for pure, small peptides	Very low resolution limits; ability to detect structural variants not presently adapted for LC/MS; long acquisition time, often >1 h
Thermospray ionization	8,000	Can be interfaced with standard-bore HPLC; can be used with NP- or RP-HPLC; simple to operate; suitable for large, polar, and involatile molecules as well as ions	Decomposition of thermolabile compounds; only volatile solvents and volatile buffers can be used; limited structural information

[a] Adapted from Ref. 53 with permission.

Table 3.8 Maximum Allowable Flow Rates for LC–Mass Spectrometry Interfaces[a]

Interface	Flow rate (μl/min)
Capillary inlet	00
Direct liquid introduction	50
Pneumatic nebulizers in medium-pressure ion sources	50
Electrospray	10
Ion spray	200
Particle beam	500
Thermospray	2000
Heated nebulizer (APCI)	2000
Atmospheric-pressure spray	2000

[a] Reprinted from Ref. 54 with permission.

neurotoxins in shellfish. High-speed drug and metabolite quantitation for drug screening in horses and analysis of drug metabolism in humans are other applications.

3.7.6 Other Detectors

A variety of other detectors have been used in HPLC. The most popular is the refractive index (RI) detector, although its use has declined during the nineties. The RI detectors are universal and nondestructive; they operate by responding to a change in the refractive index of a solvent caused by the presence of a solute in the column effluent. Such detectors are usually used in preparative LC or gel-permeation chromatography, where sensitivity is less important. In addition, they are stable and easy to operate. However, RI detectors are not suited to gradient elution, and temperature control of the column effluent and detector are critical.

Chemiluminescence is a very sensitive and selective technique. Reagent types, analytes, and detection limits have been summarized in a review by Imai.[56] Chemiluminescence has been applied to the analysis of compounds that exhibit low UV absorbance, including metal ions, amino acids, fatty acids, and bile acids. Other detectors include detectors for radioactivity, nuclear magnetic resonance (NMR), and surface-enhanced Raman spectroscopy. Radioactivity detection is one of the most selective detectors, as only components that have been radiolabeled will be detected. The interface of NMR with HPLC and has been discussed in detail by Grenier-Loustalot et al.[57] Surface-enhanced Raman spectroscopy is another technique that has been interfaced with HPLC. With a Raman spectroscopy detection, an analyte is irradiated with a monochromatic light source, and the intensity and frequency of the resultant scattered light are monitored. The scattering process involves momentary distortion of electrons distributed around a bond, followed by omnidirectional reemission as the bond returns to its original state.[25] The difference in the frequency between the incident and scattered radiation generally lies in the mid-infrared region. The advantages of the Raman detector include improved sensitivity, specificity, versatility, and detection limits compared with UV absorbance detection.

3.8 Sample Preparation

Samples rarely come in a form that can be injected directly into the instrument; some form of sample preparation usually is required. Sample preparation includes any manipulation of the sample prior to analysis, including techniques such as weighing, dilution, concentration, filtration, centrifugation, and liquid- or solid-phase extraction. Sample preparation can be performed either on-line or off-line, but it is usually performed off-line. Off-line preparation can be time-consuming and tedious, and the more steps that are required, the more susceptible the analytical method is to operator error and irreproducibility.

3.8.1 Off-line Sample Preparation Techniques

The most common off-line sample preparation techniques include filtration, concentration or dilution, weighing, pH adjustment, addition of internal standards, evaporation, centrifugation, chromatography, liquid–liquid extraction, solid-phase extraction, and derivatization.[58] Although only one of those techniques is sometimes required, usually two or more are needed to prepare the sample. Examples of samples that require pretreatment include those from the pharmaceutical industry, environmental samples where a trace amount of the analyte of interest is often masked by high concentrations of interfering substances, and physiological samples where either protein must be removed from the sample prior to analysis or else the protein itself must be isolated in native form. Most of the sample preparation techniques are straightforward, but care must be taken not to introduce contaminants. Glassware should be thoroughly cleaned, and it is often advisable to wear gloves to avoid touching the sample. Solvents and reagents should be HPLC-grade or better, and balances should be calibrated.

(i) Membrane Devices for Sample Cleanup

Membrane separation techniques, which are used mainly in industrial processes, include dialysis, electrodialysis, reverse osmosis, ultrafiltration, and microporous filtration. The major fields of applications are in water desalination, in the food, beverage, and pharmaceutical industries, and in biomedical engineering.[59] The use of membranes in analytical applications, however, is rapidly gaining in popularity, and typical applications are shown in Table 3.9.

Dialysis has been applied to the preparation of a wide range of sample types, ranging from foodstuffs to physiological fluids. Membrane-based sample preparations for chromatography have been reviewed by Van de Merbel et al.[60] In ordinary dialysis, solutes are transferred from a concentrated to a more dilute solution as a consequence of the concentration gradient.

In electrodialysis, an applied electric field rather than a concentration gradient is used to draw ions across the membrane. Because it is faster than ordinary dialysis, electrodialysis is often used in biochemical analyses for purposes of fractionation, concentration, and desalting. Reverse osmosis (RO) is a process that uses semipermeable membranes to allow water permeation; however, the membranes act as a barrier to the passage of dissolved and suspended particles. Typically, RO membranes are used to extract pure water from aqueous solutions of dissolved salts, such as seawater. The particle size cutoff is typically 0.0001 μm with driving pressures of 200 to 800 psi (1.4 to 5.5 MPa).[61]

In ultrafiltration (UF), dissolved particles, cells, and large molecules are separated from one another if they differ in size by a magnitude of 10. The size of the particles is typically 0.001 to 0.02 μm, and UF membranes are available with cutoffs for molecular species in a molecular weight range from 300 to 300,000.[61] Microporous filtration (MF) is used to clarify water and other fluids by trapping particles and microorganisms on the surface of or inside a filter. Species separated are in the range 0.1 to 10 μm.[61] The MF membranes on the smaller end of the scale find applications in gas filtration and the separation of macromolecules; the MF membranes with larger pore sizes are used for removal of particulates from materials such as hydraulic oils.

(ii) Solid-Phase Extraction Devices

Solid-phase extraction (SPE) has become an important sample preparation technique. An analyte of interest is separated from contaminants in the sample matrix by passing the gaseous or liquid sample over a solid sorbent. Various SPE devices are used both for sample concentration and to remove sample contaminants. Applications range from environmental analyses such as the detecting trace herbicides and pesticides in water and soil, to quantitations of not only drugs and drug metabolites in serum, plasma, and urine, but also vitamins and additives in feed. The devices are small cartridges or disks that contain a small quantity of sorbent either held in place by two frits, in the case of the cartridges, or enmeshed in a

Table 3.9 Summary of Membrane Processes and Some Applications

Separation process	Driving force	Separation mechanism	Species passed	Membrane structure	Application
Dialysis	Concentration gradient	Difference in diffusion rate	Ions and low MW organics	Symmetrical porous/nonporous	Separation of high and low MW compounds, water desalination
Electrodialysis	Electrical potential	Selective ion transport	Charged species (ions)	Symmetrical ionic	Water desalination, fractionation
Reverse osmosis	Pressure	Difference in membrane permeation	Water—retains virtually all ions	Symmetrical porous	Water desalination
Ultrafiltration	Pressure gradients	Difference in membrane permeation	Water and salts—retains macromolecules	Asymmetrical porous (1–100 nm)	Separation of high and low MW compounds
Microporous filtration	Pressure gradients	Sieving	Water and dissolved species	Symmetrical porous (100–1000 nm)	Bacteria filtration

network of inert fibers, in the case of the disks. The major advantage of the disks is that faster flow rates during sample loading (20–80 ml/min) may be used compared with the cartridges (1–10 ml/min), owing to the larger cross-sectional area and adsorption surface of the disks.[61] Flow rates affect purity, recovery, and sample throughput.[62] Because the disks became available only since the late 1980s, there is a greater variation of cartridge types from which to choose, and the applications literature base is significantly larger for SPE cartridges than for the disks.

There are two simple SPE strategies for sample preparation. The device is chosen so that either the analyte of interest is unretained while the interfering materials in the matrix are adsorbed, or the analyte is retained while the unwanted matrix components pass through the device. The first strategy is usually chosen when the analyte of interest is present in a high concentration. If the analyte is present at a low concentration, or if multiple components of widely differing polarities need to be isolated, the second approach is generally employed. The second approach is the usual approach for trace enrichment of extremely low-level compounds. Although SPE is a versatile technique for sample preparation, it can be labor-intensive, especially when many samples must be processed. Therefore, most manufacturers of SPE devices also sell manifolds to aid in the preparation of large numbers of samples.

Sorbents for SPE are available with normal-phase, reversed-phase or ion-exchange packings. The steps involved in SPE are illustrated in Figure 3.23.[62] The device is first conditioned by passing reagents through it to activate the functional groups on the stationary phase, followed by rinsing away the activating reagent. Once the device has been conditioned, the sample may be loaded. The device is then rinsed, and finally the analytes of interest are eluted from the device. Typical solvents and elution orders

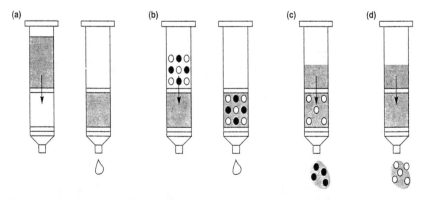

Figure 3.23. Steps in solid-phase extraction: (a) column conditioning, (b) sample loading, (c) column rinsing, and (d) elution. (Adapted from Ref. 62 with permission.)

3.8 Sample Preparation

Table 3.10 Chromatographic Modes and Elution Characteristics for Solid-Phase Extraction Devices[a]

Elution characteristic	Normal phase	Reversed phase	Ion exchange
Sorbent polarity	High	Low	High
Typical solvent polarity range	Low to medium	High to medium	High
Typical sample loading solvent	Hexane, toluene	Water, buffers	Water, buffers
Typical elution solvent	Ethyl acetate, acetone, acetonitrile	Water/ methanol, water/ acetonitrile	Buffers, salt solution
Sample elution order	Least polar components first	Most polar components first	Weakly ionized components first
Solvent change required for elution	Increase solvent polarity	Decrease solvent polarity	Increase ionic strength, change pH

[a] Reprinted from Ref. 7 with permission.

for SPE devices with the three types of packings are given in Table 3.10. The solvents must be optimized for a specific analysis, and it is possible to use other solvents. Solid-phase extraction devices may be obtained in different sizes, according to the mass of stationary phase available. The weight of stationary phase determines the sample capacity, and for particularly dirty samples total sample capacity can be easily exceeded. For most applications, however, 1 g or less of stationary phase is sufficient to retain several μg's of analyte.

(iii) Chromatography

Both column chromatography and HPLC are used routinely for sample preparation, particularly for protein samples after particulate contamination has been removed by filtration or centrifugation. In addition, the use of ultrafiltration or solid-phase extraction techniques prior to chromatography often will result in a simplified, more concentrated sample.

Gel filtration is a good technique to use at an early stage in the purification process. Because it is a gentle technique and generally will not alter or interfere with biological activity, it is particularly useful with biological samples. Gel filtration can be used to fractionate all classes of biological molecules prior to analysis. It is also useful to exchange buffers for samples obtained from ammonium sulfate precipitation.

For dilute solutions, it may be advisable to use a concentrating technique such as ion-exchange, reversed-phase, or affinity chromatography as the first step in a purification sequence. For samples containing contaminants such as cell fragments, lipids, and nucleic acid fragments that bind strongly to a column, the column will require cleaning between runs with 0.1–1.0 M NaOH. Use of such a strong base could destroy the ligand on an affinity column and would irreversibly damage a silica-based reversed-phase column. Therefore, for samples containing those types of contaminants, an ion-exchange or HIC column would be more appropriate for a first purification step.

(iv) Liquid Solvent Extraction

Liquid solvent extraction, such as Soxhlet or supercritical fluid extraction (SFE), is a technique used to extract mostly nonpolar components from solid samples. Other techniques include microwaving, Soxtec extraction, sonication and Accelerated Solvent Extraction (ASE, Dionex). In Soxhlet extraction, which is the more traditional technique, organic solvents are used to extract the compounds of interest from the sample matrix. Cooled, condensed solvents are run over the sample to extract the analytes using specialized glassware and heating elements.[63] The advantages of this technique include ease of operation, low cost, and proven performance over a wide range of sample matrices and analytes; the disadvantages include the large volumes of solvent that are required (250–500 ml per sample), the long extraction times (typically 48 hr), and the large volumes of water needed for the condensers.[63] Virtually every industry that extracts organic analytes from solid matrices has found an application for Soxhlet extraction.

In SFE supercritical carbon dioxide is usually used to extract the compounds of interest from the sample matrix, but other supercritical fluids can be used. Compounds that are ionic, extremely polar, or contain more than six hydroxyl groups, such as sugars, proteins, and phospholipids, are not soluble in carbon dioxide and therefore cannot be extracted by SFE. In SFE, the analytes are extracted using carbon dioxide at elevated pressure in an extraction chamber at elevated temperature. The pressure and temperature are chosen on the basis of the characteristics of the sample; low molecular weight compounds require a lower pressure than high molecular weight compounds. Usually, a small amount of an organic cosolvent is included to extract analytes of higher polarity and analytes that exhibit a high degree of matrix interactions.

The extracted compounds are collected into a vial that usually has in it a small volume of an organic collection solvent. The choice of collection solvent is important to ensure good recovery. Although there are many variables to consider when developing an extraction method by SFE, manufacturers of SFE instruments will provide information on the most appropriate conditions for a particular compound from a given matrix. Applications range from the extraction of fats and oils from foodstuffs to the extraction of organic compounds such as pesticides from soil samples.[64] The major advantages of SFE over the more traditional Sohxlet extractions are the increased speed of SFE (30–60 min) and the reduction in cost per analysis of SFE owing to a significant reduction in the use of organic solvents (10–30 ml per sample).[63] In addition, the extraction solvent is more easily removed from the analytes than in Soxhlet extraction.[63] The instrumentation required for SFE, however, is considerably more expensive than that for Soxhlet extraction.[63]

Microwave extraction uses microwaves to heat solvents that are placed in the sample container. The microwaves heat the solvent and also add pressure to the sample vessel. The solvent is then removed and treated. The advantages of the microwave technique over Soxhlet extraction include reduced solvent consumption (~40 ml per sample), rapid extractions (12 samples/hr), and apparent ease of use.[63] The disadvantages include the initial capital investment, lack of approval from government agencies, and the need to use solvents that are microwave compatible.[63] Applications include those in the food and environmental fields.

The most recent of the liquid solvent extraction techniques involves a proprietary solvent extraction device called ASE (for Accelerated Solvent Extraction). The ASE device uses a combination of temperature, pressure, and standard organic solvents for extracting organic materials from solid samples.[63] Temperatures range from 50 to 200°C, and pressure is applied up to a limit of 3000 psi (21 MPa). The advantages of ASE include speed, typically averaging 15 min per sample, ease of use, and minimal use of solvents (13–15 ml per 10-g sample size); ASE also meets the requirements of U.S. Environmental Protection Agency (EPA) Method 3545 (proposed).[63] Disadvantages include the initial capital investment and the low number of applications that have been developed as a result of the recent (1995) appearance of ASE in the marketplace.[63] Nevertheless, ASE has been applied to the extraction of PAHs;[65] polychlorinated biphenyls (PCBs);[65] chlorinated herbicides[65] and pesticides;[65] organophosphorus pesticides;[65] bases, neutrals, and acids;[65] as well as dioxins and furans[66] in environmental samples such as bay sediment, chimney brick, urban dust, and fish tissue, from food samples such as cookies,[65] dog biscuits,[65] cereal,[65] potato chips,[65] fruits, and vegetables, and from pharmaceuticals such as animal feeds and vitamins.[63]

3.8.2 On-line Sample Preparation Techniques

(i) Autosamplers

Autosamplers are used to inject samples automatically. They tend to be used for routine analyses, after a chromatographic method has been developed and validated. Some autosamplers can be programmed to perform dilutions and standard additions prior to autoinjection, and these instruments can be useful during methods development. Autosamplers can improve sample recoveries and method reproducibility by reducing the amount of hands-on sample preparation performed by the analyst.

(ii) Solid-Phase Extraction Devices

The most simple of the on-line solid-phase extraction devices is the guard column. Guard columns are smaller versions of the analytical column that are placed before the analytical column to protect the analytical column from contamination. Because the packing inside the guard column is the same as the packing inside the analytical column, the same analytes will adhere to the guard as would adhere to the analytical column. By placing a guard column in front of the analytical column, therefore, any species present in the sample that would bind tightly or irreversibly to the analytical column will be trapped by the guard column and will never reach the analytical column. The advantage of using a guard column is the considerably lower expense of replacing a contaminated guard column compared to an analytical column.

The other form of on-line solid-phase extraction procedures involves column-switching techniques. Column switching employs valves that can be switched manually or automatically between a number of columns at predetermined times.[67-69] For sample cleanup the analyte of interest is retained on the primary or precolumn while the interfering matrix components are eluted to waste. The analytes are then diverted to a second or analytical column where they are separated for identification and quantification.

Column-switching techniques can be used to concentrate a dilute sample for trace analysis or to isolate a particular group of analytes from the rest of the sample matrix. When the technique is used to concentrate a group of analytes, the precolumn is known as a concentrator column. The concentrator column usually contains the same stationary phase as the analytical column and is often simply a guard column.[70] A typical column-switching valve configuration is illustrated in Figure 3.24. In the load position, the eluent is pumped through the analytical column via a bypass loop. The sample is loaded onto the concentrator column, and the analytes are concentrated at the head while the rest of the sample matrix is sent to waste. When sufficient sample has been loaded, the valve is switched so that the eluent is diverted through the concentrator column on its way to the analytical column. The analytes are eluted from the head of the concentrator column in the opposite direction from that in which they were loaded, so that no further dilution of the sample occurs.

When column switching is used for sample cleanup, the technique is known as zone cutting. If the fraction of effluent to be transferred from the precolumn to the analytical column is at the front of the precolumn chromatogram, the technique is known as front cutting. Heart-cut technique is the name used when the fraction to be transferred to the analytical column is in the middle of the precolumn chromatogram, whereas end-cut technique is the name given if the fraction to be transferred is at the tail end of the precolumn chromatogram. The dimensions of the precolumn should be chosen so that the column has a high loading capacity. The high loading capacity prevents losses of the analytes by breakthrough while the sample matrix is being flushed to waste. The particle size of the precolumn should be large (10–40 μm) compared with those in the analytical column to prevent clogging problems.[64] The retention capability of the precolumn should be lower than that of the analytical column so that the analytes will be concentrated at the head of the analytical column prior to the elution. The advantages and disadvantages of column-switching techniques for sample cleanup are given in Table 3.11.

3.9 Troubleshooting

Most analysts will be required to troubleshoot the chromatography system at some stage. Although chromatographic instruments are considerably more rugged now than they once were, components still have limited lifetimes and need to be replaced. As operator error can be the cause of a problem, it is important to understand the operation of the system in order to minimize downtime. The best approach to troubleshooting, however, is preventive maintenance: the most important points to remember are filter, degas, and flush.[71]

Filtration applies not only to the mobile phase and sample, but also to the chromatographic system itself in the form of in-line filters and guard columns. The main purpose of filtration is to remove large particulates that could plug up the system. When a pure mobile phase such as HPLC-grade water or organic solvent is used, filtration is not so important, as the liquids

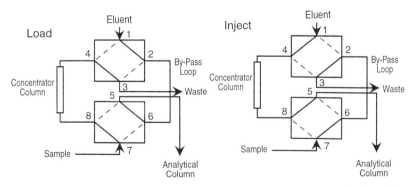

Figure 3.24 Configuration of a four-way valve for sample preconcentration. (Adapted from Ref. 70 with permission.)

Table 3.11 Advantages and Disadvantages of Column-Switching Techniques for Sample Cleanup[a]

Advantages	Disadvantages
Minimal sample handling	Requirement for switching valves and additional columns and/or pumping systems
On-line sample processing	Requirement for compatible mobile phases
Considerable time reduction	Need for periodic replacement of precolumn
Possibility of full automation	
Greater precision and accuracy	
Improvement in selectivity by combining different chromatographic modes	
Minor consumption of organic solvents	
Protection of photolabile analytes	

[a] Adapted from Ref. 67 with permission.

are filtered prior to bottling. However, any mixture of liquids should be filtered through a 0.5-μm (or smaller) porosity filter. The mobile phase should always be filtered and the glassware or plasticware be scrupulously clean so that other contaminants are not introduced. The solvent lines leading from the solvent reservoirs into the pump should also have inlet-line filters in place to filter out particles and to hold the inlet line at the bottom of the reservoir so that no air is drawn into the pump. Finally, the sample should be filtered prior to injection if it is cloudy or contains particulate matter; otherwise, it can damage the injection valve or block the column inlet frit. Guard columns are recommended because they act both as physical filters when they collect particulate matter on the frits and as chemical filters when they remove sample contaminants that could adversely affect the chromatography.

Degassing the mobile phase prior to introduction into the chromatographic system is one of the easiest ways to improve system reliability. If the mobile phase is not degassed, trapped air is likely to come out of solution under pressure. When this happens, the pump will lose prime. The most common way to degas the mobile phase is either by sparging the reservoir with helium or by vacuum degassing using a vacuum pump. The second method can be performed at the same time as the mobile phase is being filtered, if the sample is drawn through the filter by vacuum.

The final step in preventative maintenance is to flush the system on a regular basis to wash out any contaminants that have accumulated in the system over time. If the mobile phase is purely aqueous (i.e., buffers but

3.9 Troubleshooting

no organic solvents), then the system can be flushed with water. If it contains an organic solvent, the first step is to remove any salt components that could precipitate under strong solvent conditions. The eluent strength is then increased to flush out any contaminants. To remove the salt components, the mobile phase is switched from solvent–buffer to the same proportion of solvent–water, and approximately five column volumes of this new mixture is pumped through the system.[71] The strength of the organic component is then increased to finish the flushing process. If the mobile phase contains a high concentration of salt, the salt will invariably accelerate the deterioration of the pump seals. Most commercial pumps are equipped with a flushing channel on each pump head, and it is advisable to flush about 10 ml of warm water through the channel to remove any buffer leakage. If flushing is performed on a regular basis, the pump seals will last indefinitely.

If, despite these precautions, the system fails to perform as expected, the first step in troubleshooting is to define the problem. The presence of a problem is usually apparent from the chromatogram and can be manifested in any number of ways: no peaks, poor peak shape, poor sensitivity, baseline drift, spikes in the baseline, drift in retention times, resolution deterioration, etc. Most instrument manufacturers[71–73] have literature addressing the more common causes of these effects, and the books by Snyder and Kirkland[2] and by Dolan and Snyder[74] can be consulted for more detailed help. In this section, we will assume that a method has been developed and has been running well and the problem has recently manifested itself. Thus, problems such as poor peak shapes that result from poorly cut tubing are not considered here. Table 3.12 lists some of the more commonly encountered problems, probable causes, and possible remedies.[73]

Once the problem has been identified, a list of probable causes should be made. For example, if the problem is a sudden high backpressure, possible causes are a blocked column frit, sample precipitation on the column, changed or incorrect flow rate, blocked tubing, defective pressure transducer, or even a drop in the operating temperature. When the list has been made, it is time to investigate the cause, but it is important to remember to change just one variable at a time, so that cause and effect can be determined. The parts of the system that can be changed during troubleshooting are the instrument settings, the column, the sample, the mobile phase, and malfunctioning instrument components. Always start with the simple things and then progress to more complicated things. For example, if no peaks are visible, first check to see if there is eluent in the reservoir, if the sample has not degraded, if the detector is switched on, etc., and then check for malfunctioning of the detector. Finally, if the system continues to malfunction, call the instrument manufacturer. Most manufacturers have telephone support personnel who are very knowledgable and helpful. As a last resort, call in the service representative for the instrument manufac-

Table 3.12 HPLC Problems, Probable Causes, and Remedies[a]

Problem	Probable cause	Remedy/Comments
Problem No. 1: No Peaks/Very Small Peaks		
Normal / Problem / Problem	1. Detector lamp off. 2. Loose/broken wire between detector and integrator or recorder. 3. No mobile phase flow. 4. No sample/deteriorated sample. 5. Settings too high on detector or recorder.	1. Turn lamp on. 2. Check electrical connections and cables. 3. See "No Flow" (Problem No. 2). 4. Be sure automatic sampler vials have sufficient liquid. Evaluate system performance with fresh standard to confirm sample as source of problem. 5. Check attenuation or gain settings.

Problem	Probable cause	Remedy/Comments
Problem No. 2: No Flow		
Normal / Problem	1. Pump off. 2. Flow interrupted/obstructed. 3. Leak. 4. Air trapped in pump head. (Revealed by pressure fluctuations.)	1. Start pump. 2. Check mobile phase level in reservoir(s). Check flow throughout system. Examine sample loop for obstruction or air lock. Make sure mobile phase inlet filter is clean. Make sure mobile phase components are miscible and mobile phase is properly degassed. 3. Check system for loose fittings. Check pump for leaks, salt buildup, unusual noises. Change pump seals if necessary. 4. Disconnect tubing at guard column (if present) or analytical column inlet. Check for flow. Purge pump at high flow rate (10 ml/min), prime system if necessary. (Prime each pump head separately.) If system has check valve, loosen valve to allow air to escape. If problem persists, flush system with 100% methanol. If problem still persists, contact system manufacturer.
Problem No. 3: No Pressure/Pressure Lower than Usual		
Normal / Problem	1. Leak. 2. Mobile phase flow interrupted/obstructed. 3. Air trapped in pump head. (Revealed by pressure fluctuations.) 4. Leak at column inlet end fitting. 5. Air trapped elsewhere in system. 6. Worn pump seal causing leaks around pump head.	1. Check system for loose fittings. Check pump for leaks, salt buildup, unusual noises. Change pump seals if necessary. 2. Check mobile phase level in reservoir(s). Check flow throughout system. Examine sample loop for obstruction or air lock. Make sure mobile phase inlet filter is clean. Make sure mobile phase components are miscible and mobile phase is properly degassed. 3. Disconnect tubing at guard column (if present) or analytical column inlet. Check for flow. Purge pump at high flow rate (10 ml/min), prime system if necessary. (Prime each pump head separately.) If system has check valve, loosen valve to allow air to escape. 4. Reconnect column and pump solvent at 1–2 ml/min. If pressure is still low, check for leaks at inlet fitting or column end fitting. 5. Disconnect guard and analytical column and purge system. Reconnect column(s). If problem persists, flush system with 100% methanol. 6. Replace seal.

Problem	Probable cause	Remedy/Comments
Problem No. 4: Pressure Higher than Usual		
Normal / Problem	1. Problem in pump, injector, or tubing.	1. Disconnect guard and analytical column, run pump at 2–5 ml/min. If pressure is minimal, see Cause 2. If not, isolate cause by systematically eliminating system components, starting with detector and working back to pump.
	2. Obstructed guard or analytical column.	2. Remove guard column (if present) and check pressure. Replace guard column if necessary. If analytical column is obstructed, reverse and flush. If problem persists, column may be clogged with strongly retained contaminants. Use appropriate restoration procedure.
Problem No. 5: Variable Retention Times		
Normal	1. Leak.	1. Check system for loose fittings. Check pump for leaks, salt buildup, unusual noises. Change pump seals if necessary.
	2. Change in mobile phase composition. (Small changes can lead to large changes in retention times.)	2. Check makeup of mobile phase. If mobile phase is machine mixed, hand mix and supply from one reservoir.
	3. Air trapped in pump. (Retention times increase and decrease at random times.)	3. Purge air from pump head or check valves. Change pump seals if necessary. Be sure mobile phase is degassed.
Problem	4. Column temperature fluctuations (especially evident in ion-exchange systems).	4. Use reliable column oven or insulate column. (**Note:** higher column temperatures increase column efficiency. For optimum results, head eluant before introducing it onto column.)
	5. Column overloading. (Retention times usually decrease as mass of solute injected on column exceeds column capacity.)	5. Inject smaller volume (e.g., 10 μl vs. 100 μl) or make 1:10 and 1:100 dilutions of sample.
	6. Sample solvent incompatible with mobile phase.	6. Adjust solvent. Whenever possible, inject samples in mobile phase.
Problem	7. Column problem. (Not a common cause of erratic retention. As a column ages, retention times gradually decrease.)	7. Substitute new column of same type to confirm column as cause. Discard old column if restoration procedures fail.

3.9 Troubleshooting 123

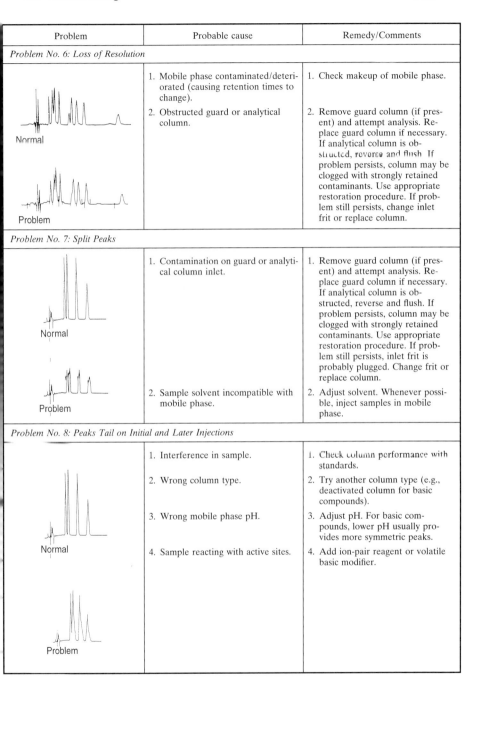

Problem	Probable cause	Remedy/Comments
Problem No. 6: Loss of Resolution		
	1. Mobile phase contaminated/deteriorated (causing retention times to change).	1. Check makeup of mobile phase.
	2. Obstructed guard or analytical column.	2. Remove guard column (if present) and attempt analysis. Replace guard column if necessary. If analytical column is obstructed, reverse and flush. If problem persists, column may be clogged with strongly retained contaminants. Use appropriate restoration procedure. If problem still persists, change inlet frit or replace column.
Problem No. 7: Split Peaks		
	1. Contamination on guard or analytical column inlet.	1. Remove guard column (if present) and attempt analysis. Replace guard column if necessary. If analytical column is obstructed, reverse and flush. If problem persists, column may be clogged with strongly retained contaminants. Use appropriate restoration procedure. If problem still persists, inlet frit is probably plugged. Change frit or replace column.
	2. Sample solvent incompatible with mobile phase.	2. Adjust solvent. Whenever possible, inject samples in mobile phase.
Problem No. 8: Peaks Tail on Initial and Later Injections		
	1. Interference in sample.	1. Check column performance with standards.
	2. Wrong column type.	2. Try another column type (e.g., deactivated column for basic compounds).
	3. Wrong mobile phase pH.	3. Adjust pH. For basic compounds, lower pH usually provides more symmetric peaks.
	4. Sample reacting with active sites.	4. Add ion-pair reagent or volatile basic modifier.

124   Chapter 3 Instrumentation for High-Performance Liquid Chromatography

Problem	Probable cause	Remedy/Comments
Problem No. 9: Peaks Formerly Symmetric, Now Tailing		
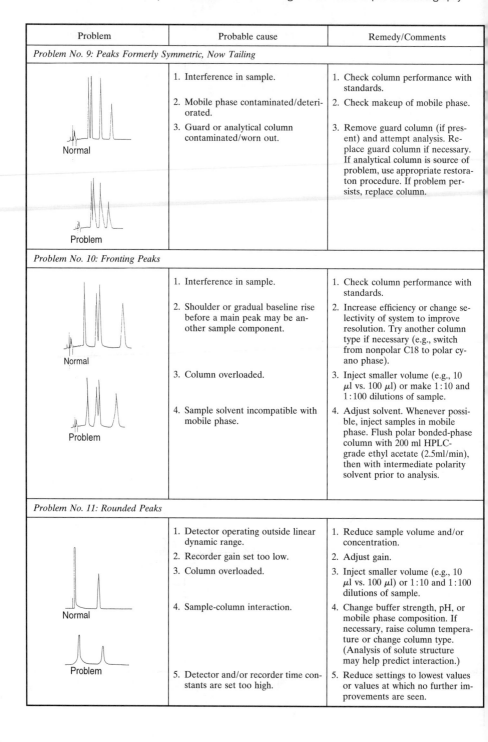	1. Interference in sample.	1. Check column performance with standards.
	2. Mobile phase contaminated/deteriorated.	2. Check makeup of mobile phase.
	3. Guard or analytical column contaminated/worn out.	3. Remove guard column (if present) and attempt analysis. Replace guard column if necessary. If analytical column is source of problem, use appropriate restoraton procedure. If problem persists, replace column.
Problem No. 10: Fronting Peaks		
	1. Interference in sample.	1. Check column performance with standards.
	2. Shoulder or gradual baseline rise before a main peak may be another sample component.	2. Increase efficiency or change selectivity of system to improve resolution. Try another column type if necessary (e.g., switch from nonpolar C18 to polar cyano phase).
	3. Column overloaded.	3. Inject smaller volume (e.g., 10 μl vs. 100 μl) or make 1:10 and 1:100 dilutions of sample.
	4. Sample solvent incompatible with mobile phase.	4. Adjust solvent. Whenever possible, inject samples in mobile phase. Flush polar bonded-phase column with 200 ml HPLC-grade ethyl acetate (2.5ml/min), then with intermediate polarity solvent prior to analysis.
Problem No. 11: Rounded Peaks		
	1. Detector operating outside linear dynamic range.	1. Reduce sample volume and/or concentration.
	2. Recorder gain set too low.	2. Adjust gain.
	3. Column overloaded.	3. Inject smaller volume (e.g., 10 μl vs. 100 μl) or 1:10 and 1:100 dilutions of sample.
	4. Sample-column interaction.	4. Change buffer strength, pH, or mobile phase composition. If necessary, raise column temperature or change column type. (Analysis of solute structure may help predict interaction.)
	5. Detector and/or recorder time constants are set too high.	5. Reduce settings to lowest values or values at which no further improvements are seen.

Problem	Probable cause	Remedy/Comments
Problem No. 12: Baseline Drift		
Normal Problem	1. Column temperature fluctuation. (Even small changes cause cyclic baseline rise and fall. Most often affects refractive index and conductivity detectors. UV detectors at high sensitivity or in indirect photometric mode.) 2. Nonhomogeneous mobile phase. (Drift usually to higher absorbance, rather than cyclic pattern from temperature fluctuation.) 3. Contaminant or air buildup in detector cell. 4. Plugged outlet line after detector. (High pressure cracks cell window, producing noisy baseline.) 5. Mobile phase mixing problem or change in flow rate. 6. Slow column equilibration, especially when changing mobile phase. 7. Mobile phase contaminated, deteriorated, or prepared from low quality materials. 8. Strongly retained materials in sample (high k') can elute as very broad peaks and appear to be a rising baseline. (Gradient analyses can aggravate problem.) 9. Mobile phase recycled but detector not adjusted. 10. Detector (UV) not set at absorbance maximum but at slope of curve.	1. Control column and mobile phase temperature, use heat exchanger before detector. 2. Use HPLC-grade solvents, high purity salts, and additives. Degas mobile phase before use, sparge with helium during use. 3. Flush cell with methanol or other strong solvent. If necessary, clean cell with 1N HNO_3 (*never* with HCl). 4. Unplug or replace line. Refer to detector manual to replace window. 5. Correct composition/flow rate. To avoid problem, routinely monitor composition and flow rate. 6. Flush column with intermediate strength solvent, run 10–20 column volumes of new mobile phase through column before analysis. 7. Check makeup of mobile phase. 8. Use guard column. If necessary, flush column with strong solvent between injections or periodically during analysis. 9. Reset baseline. Use new materials when dynamic range of detector is exceeded. 10. Change wavelength to UV absorbance maximum.
Problem No. 13: Baseline Noise (regular)		
Normal Problem	1. Air in mobile phase, detector cell, or pump. 2. Leak. 3. Incomplete mobile phase mixing. 4. Temperature effect (column at high temperature, detector unheated). 5. Other electronic equipment on same line. 6. Pump pulsations.	1. Degas mobile phase. Flush system to remove air from detector cell or pump. 2. Check system for loose fittings. Check pump for leaks, salt buildup, unusual noises. Change pump seals if necessary. 3. Mix mobile phase by hand or use less viscous solvent. 4. Reduce differential or add heat exchanger. 5. Isolate LC, detector, recorder to determine if source of problem is external. Correct as necessary. 6. Incorporate pulse dampener into system.

Problem	Probable cause	Remedy/Comments
Problem No. 14: Baseline Noise (irregular)		
Normal Problem	1. Leak	1. Check system for loose fittings. Check pump for leaks, salt buildup, unusual noises. Change pump seals if necessary.
	2. Mobile phase contaminated, deteriorated, or prepared from low quality materials.	2. Check makeup of mobile phase.
	3. Detector/recorder electronics.	3. Isolate detector and recorder electronically. Refer to instruction manual to correct problem.
	4. Air trapped in system.	4. Flush system with strong solvent.
	5. Air bubbles in detector.	5. Purge detector. Install backpressure device after detector.
	6. Detector cell contaminated. (Even small amounts of contaminants can cause noise.)	6. Clean cell.
	7. Weak detector lamp.	7. Replace lamp.
	8. Column leaking silica or packing material.	8. Replace column.
Problem No. 15: Broad Peaks		
Normal Problem	1. Mobile phase composition changed.	1. Prepare new mobile phase.
	2. Mobile phase flow rate too low.	2. Adjust flow rate.
	3. Leak (especially between column and detector).	3. Check system for loose fittings. Check pump for leaks, salt buildup, and unusual noises. Change pump seals if necessary.
	4. Detector settings incorrect.	4. Adjust settings.

	5. Extra-column effects:
	  a. Column overloaded.
	
	  b. Detector response time or cell volume too large.
	  c. Tubing between column and detector too long or ID too large.
	  d. Recorder response time too high.
		5.
		  a. Inject smaller volume (e.g., 10 μl vs. 100 μl) or make 1:10 and 1:100 dilutions of sample.
		  b. Reduce response time or use smaller cell.
		  c. Use as short a piece of 0.007–0.010″ ID tubing as practical.
		  d. Reduce response time.

	6. Buffer concentration too low.	6. Increase concentration.
	7. Guard column contaminated/worn out.	7. Replace guard column.
	8. Column contaminated/worn out.	8. Replace column with new one of same type. If new column provides symmetrical peaks, flush old column.
	9. Void at column inlet.	9. Open inlet end and fill void or replace column.
	10. Peak represents two or more poorly resolved compounds.	10. Change column type to improve separation.
	11. Column temperature too low.	11. Increase temperature. Do not exceed 75°C unless higher temperatures are acceptable to column manufacturer.

3.9 Troubleshooting

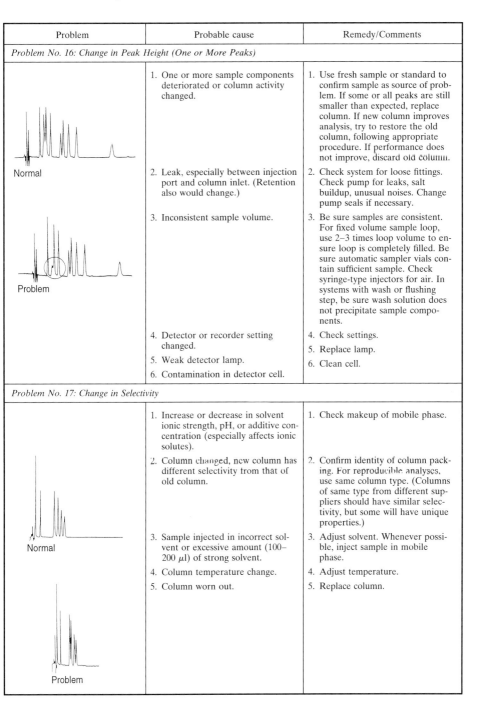

Problem	Probable cause	Remedy/Comments
Problem No. 16: Change in Peak Height (One or More Peaks)		
Normal / Problem	1. One or more sample components deteriorated or column activity changed.	1. Use fresh sample or standard to confirm sample as source of problem. If some or all peaks are still smaller than expected, replace column. If new column improves analysis, try to restore the old column, following appropriate procedure. If performance does not improve, discard old column.
	2. Leak, especially between injection port and column inlet. (Retention also would change.)	2. Check system for loose fittings. Check pump for leaks, salt buildup, unusual noises. Change pump seals if necessary.
	3. Inconsistent sample volume.	3. Be sure samples are consistent. For fixed volume sample loop, use 2–3 times loop volume to ensure loop is completely filled. Be sure automatic sampler vials contain sufficient sample. Check syringe-type injectors for air. In systems with wash or flushing step, be sure wash solution does not precipitate sample components.
	4. Detector or recorder setting changed.	4. Check settings.
	5. Weak detector lamp.	5. Replace lamp.
	6. Contamination in detector cell.	6. Clean cell.
Problem No. 17: Change in Selectivity		
Normal / Problem	1. Increase or decrease in solvent ionic strength, pH, or additive concentration (especially affects ionic solutes).	1. Check makeup of mobile phase.
	2. Column changed, new column has different selectivity from that of old column.	2. Confirm identity of column packing. For reproducible analyses, use same column type. (Columns of same type from different suppliers should have similar selectivity, but some will have unique properties.)
	3. Sample injected in incorrect solvent or excessive amount (100–200 μl) of strong solvent.	3. Adjust solvent. Whenever possible, inject sample in mobile phase.
	4. Column temperature change.	4. Adjust temperature.
	5. Column worn out.	5. Replace column.

Problem	Probable cause	Remedy/Comments
Less Serious/Significant Problems		
Problem No. 18: Negative Peak		
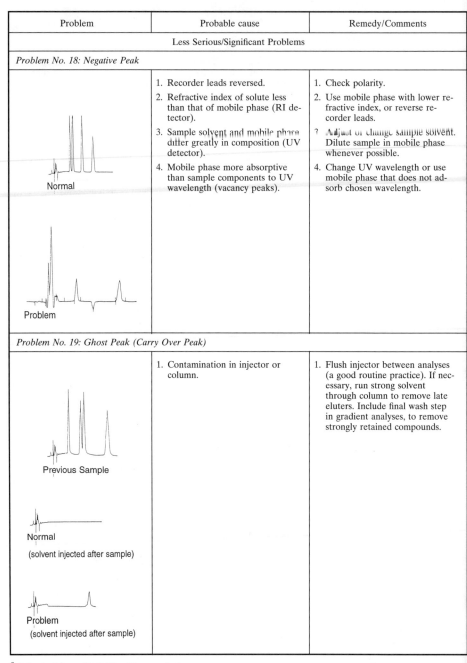	1. Recorder leads reversed. 2. Refractive index of solute less than that of mobile phase (RI detector). 3. Sample solvent and mobile phase differ greatly in composition (UV detector). 4. Mobile phase more absorptive than sample components to UV wavelength (vacancy peaks).	1. Check polarity. 2. Use mobile phase with lower refractive index, or reverse recorder leads. 3. Adjust or change sample solvent. Dilute sample in mobile phase whenever possible. 4. Change UV wavelength or use mobile phase that does not adsorb chosen wavelength.
Problem No. 19: Ghost Peak (Carry Over Peak)		
	1. Contamination in injector or column.	1. Flush injector between analyses (a good routine practice). If necessary, run strong solvent through column to remove late eluters. Include final wash step in gradient analyses, to remove strongly retained compounds.

[a] Adapted from Ref. 73 with permission.

turer. If the service representative is busy and unable to help immediately, the system will be down and time will be wasted.

3.10 Summary of Major Concepts

1. An HPLC instrument consists of

eluent containers,
pump,
injection device,
column,
detector,
waste container, and
data station.

The pump, injector, column, and detector are connected together with narrow-inner-diameter tubing, to minimize band broadening.

2. Pumps can be categorized in three different ways, according to

flow rate,
materials of construction, and
mechanism of eluent displacement.

3. Pump pulsations are minimized by the use of

noncircular cam to drive the piston and
pulse dampers.

4. Flow reproducibility is optimized by the use of

pressure loops and
flow loops.

5. Solvent degassing is important to ensure that air bubbles do not outgas and interfere with accurate delivery of flow. Solvents can be degassed by

external vacuum degassing,
helium sparging, and
on-line degassing.

6. Pumping systems are designed to deliver either a single eluent or multiple eluents. These are known as

isocratic pumps and
gradient pumps.

Blending of two or more eluents by a gradient pump is achieved in one of two ways

high-pressure mixing and
low-pressure mixing.

7. The column is the heart of the chromatograph, providing the means for separating a mixture into components. The length, diameter, and construction material of the column affect the lifetime, efficiency, and speed of separation. The size and nature of the packing material affect resolution.

8. The detector converts a change in column effluent into an electrical signal that is recorded by the data station. Detectors in HPLC include

UV absorbance,
fluorescence,
electrochemical,
mass spectrometry, and
others including refractive index, nuclear magnetic resonance, chemiluminescence, radioactivity, and surface-enhanced Raman spectroscopy.

9. Sample preparation includes any manipulation of the sample prior to analysis, including such techniques as weighing, dilution, concentration, filtration, centrifugation, and liquid- or solid-phase extraction. These techniques may be performed with the aid of devices such as

membrane devices,
on-line and off-line solid-phase extraction devices,
chromatography,
liquid solvent extraction, and
autosamplers.

References

1. Poole, C. F., and Poole, S. K., "Chromatography Today." Elsevier, New York, 1991.
2. Snyder, L. R., and Kirkland, J. J., "Introduction to Modern Liquid Chromatography," 2nd Ed. Wiley, New York, 1979.
3. Hupe, K.-P., Rozing, G., and Schrenker, H., "High Performance Liquid Chromatography in Biochemistry", (A. Henschen, K.-P. Hupe, F. Lottspeich, and W. Voelter, eds.), VCH Verlagsgesellschaft mbH, D-6940 Weinheim, 1985.
4. Spectrophysics Inc., Mountain View, CA, U.S. Patent #4,133,767.
5. Proposed standard practice for the evaluation of gradient HPLC pumps, ASTM E-19.09.07, September, 1989.
6. Spruce, B., and Bakalyar, S. R., "Troubleshooting Guide for HPLC Injection Problems," 2nd Ed. Rheodyne Incorporated, Cotati, California, 1992.
7. "The Waters Chromatography Handbook," P/C WAT056930. Millipore Corporation, Milford, Massachusetts, 1993–1994.

8. Unger, K. K., and Trudinger, U., *in* "High Performance Liquid Chromatography," (P. R. Brown and R. A. Hartwick, eds.), Chapter 3. Wiley, New York, 1989.
9. Cserhati, T., *Chromatographia* **29**, 593 (1990).
10. Lingeman, H., van Munster, H. A., Beyben, J. H., Underber, W. J. M. and Hulshoff, A., *J. Chromatogr.* **352**, 261 (1986).
11. Hanggi, D. A., and Marks, N. R., *LC-GC.* **11**, 128 (1993).
12. Knox, J. H., and Kauer, B., *in* "High Performance Liquid Chromatography," (P. R. Brown and R. A. Hartwick, eds.), Chapter 4. Wiley, New York, 1989.
13. Kriz, J., Adamcova, E., Knox, J. H., and Hora, J., *J. Chromatogr.* **663**, 151 (1994).
14. Hiranuma, T., Horigome, T., and Sugano, H., *J. Chromatogr.* **515**, 399 (1990).
15. Lundahl, P., Watanabe, Y., and Takagi, T., *J. Chromatogr.* **604**, 95 (1992).
16. Pietrzyk, D. J., *in* "High Performance Liquid Chromatography," (P. R. Brown and R. A. Hartwick, eds.), Chapter 5. Wiley, New York, 1989.
17. Haddad, P. R., and Jackson, P. E., *J. Chromatogr. Libr. Ser.* **46**, 30 (1990).
18. Sadek, P. D., Carr, P. W., Bowers, L. D., and Haddad, L. C., *Anal. Biochem.* **144**, 128 (1985).
19. Silverstein, R. M., Bassler, G. C., and Morrill, T. C., *in* "Spectrometric Identification of Organic Compounds," 4th Ed., Chapter 6. p. 312, Wiley, New York, 1981.
20. Canova-Davis, E., Baldonado, I. P., and Teshima, G. M., *J. Chromatogr.* **508**, 81 (1990).
21. Li, J. B., *LC-GC.* **10**, 856 (1992).
22. Yeung, E. S., *Acc. Chem. Res.* **22**, 125 (1989).
23. Schill, G., and Arvidsson, E., *J. Chromatogr. Biomed. Applic.* **492**, 299 (1989).
24. Denkert, M., Hackzell, L., Schill, G., and Sjorgren, E., *J. Chromatogr.* **218**, 31 (1981).
25. Crommen, J., *J. Pharm. Biomed. Anal.* **1**, 549 (1983).
26. Hackzell, L., and Schill, G., *Chromatographia* **15**, 437 (1982).
27. Herne, P., Renson, M., and Crommen, J., *Chromatographia* **19**, 274 (1984).
28. Takeuchi, T., and Ishi, D., *J. Chromatogr.* **408**, 324 (1988).
29. Takeuchi, T., and Ishi, D., *J. Chromatogr.* **396**, 149 (1987).
30. Johnson, E. L., and Stevenson, R., "Basic Liquid Chromatography." Varian Associates, Palo Alto, California, 1978.
31. Larew, L. A., and Johnson, D. C., *Anal. Chem.* **60**, 1867 (1988).
32. Koerner, P. J., and Nieman, T. A., *J. Chromatogr.* **499**, 217 (1988).
33. Weinberger, R., "Practical Capillary Electrophoresis," Chapter 10. Academic Press, San Diego, California, 1993.
34. Fuerst, P., Pollack, L., Graser, T. A., Godel, H., and Stehle, P., *J. Chromatogr.* **499**, 577 (1990).
35. Lunte, S. M., and Wong, O. S., *Curr. Sep.* **10**, 19 (1990).
36. Coquet, A., Veuthey, J. L., and Haerdi, W., *J. Chromatogr.* **553**, 255 (1991).
37. Gatti, R., Cavrini, V., Roveri, P., and Pinzauti, S., *J. Chromatogr.* **507**, 451 (1990).

38. Toyooka, T., Ishibashi, M., Takeda, Y., and Imai, K., *Analyst (London)* **116**, 609 (1991).
39. Nozaki, O., Ohata, T., Ohba, Y., Moriyama, H., and Kato, Y., *J. Chromatogr.* **570**, 1 (1991).
40. Tsuruta, Y., Date, Y., and Kohashi, K., *Anal. Sci.* **7**, 411 (1991).
41. Roemen, T. M., and Van der Vusse, G. J., *J. Chromatogr.* **570**, 243 (1990).
42. Shijo, Y., Tanaka, K., and Uehara, N., *Anal. Sci.* **7**, 507 (1991).
43. Van der Horst, F. A. L., Teeuwsen, J., Holthius, J. J. M., and Brikman, U. A. T., *J. Pharm. Biomed. Anal.* **7**, 799 (1990).
44. Pickering, M. V., *LC-GC* **6**, 994 (1988).
45. Boppana, V. K., and Rhodes, G. R., *J. Chromatogr.* **507**, 799 (1990).
46. Yonekura, S., Iwasaki, M., Kai, M., and Ohkura, Y., *J. Chromatogr.* **654**, 19 (1994).
47. Wildespin, A. F., and Green, N. M., *Anal. Biochem.* **132**, 449 (1983).
48. Sauda, K., Imasaka, T., and Ishibashi, N., *Anal. Chem.* **58**, 2649 (1986).
49. Rocklin, R., "Conductivity and Amperometry: Electrochemical Detection in Liquid Chromatography." Dionex Corporation, Sunnyvale, California, 1989.
50. Haddad, P. R., and Heckenberg, A. L., *J. Chromatogr.* **300**, 357 (1984).
51. Haddad, P. R., and Foley, R. C., *Anal. Chem.* **61**, 1435 (1989).
52. Small, H., Stevens, T. S., and Bauman, W. C., *Anal. Chem.* **47**, 1801 (1975).
53. Carr, S. A., Hemling, M. E., Bean, M. F., and Roberts, G. D., *Anal. Chem.* **63**, 2802 (1991).
54. Niessen, W. M. A., Tjaden, U. R., and van der Greef, J., *J. Chromatogr.* **554**, 3 (1991).
55. Allen, M. H., and Shushan, B. I., *LC-GC* **11**, 112 (1993).
56. Imai, K., *Chromatogr. Sci.* **48**, 359 (1990).
57. Grenier-Loustalot, M. F., Grenier, P., Bounoure, J., Grall, M., and Panaras, R., *Analusis* **18**, 200 (1990).
58. Majors, R. E., *LC-GC* **10**, 912 (1992).
59. Strathmann, H., *Swiss Biotechnol.* **7**, 13 (1989).
60. van de Merbel, N. C., Hageman, J. J., and Brinkman, U. A. T., *J. Chromatogr.* **634**, 1 (1993).
61. Snyder, J. L., Grob, R. L., McNally, M. E., and Oostdyk, T. S., *LC-GC* **12**, 230 (1994).
62. Jordan, L., *LC-GC* **11**, 634 (1993).
63. Dionex New Product Introduction Manual, Pittsburgh Conference. Dionex Corporation, Sunnyvale, California, 1995.
64. Lopez-Avila, V., Dodhiwala, N. S., Benedicto, J., and Beckert, W. F., "Supercritical Fluid Extraction of Organic Compounds from Standard Reference Materials," EPA Report 600/X-91/149. U.S. Environmental Protection Agency, Washington, D.C., 1991.
65. Dionex Application Notes 313, 316, 317, 318, 319, 320, and 321, Dionex Corporation, Sunnyvale, California.
66. Knowles, D., *Chemosphere* **March** (1996). In press.
67. Campins-Falco, P., Herraez-Hernandez, R., and Sevillano-Cabeza, A., *J. Chromatogr.* **619**, 177 (1993).
68. Ramsteiner, K. A., *J. Chromatogr.* **456**, 3 (1988).
69. Koenigbauer, M. J., and Majors, R. E., *LC-GC* **8**, 512 (1990).

70. "The Use of Concentrator Columns in Ion Chromatography," Technical Note 8, LPN 0576 3M 8/94. Dionex Corporation, Sunnyvale, California, 1994.
71. Dolan, J. W., *LC-GC* **10**, 842 (1992).
72. "Fundamentals of Maintenance and Troubleshooting." Dionex Corporation Training Manual, Dionex Corporation, Sunnyvale, California, 1992.
73. "HPLC Troubleshooting Guide," Guide 826. Supelco Division of Rohm and Haas, Bellefonte, Pennsylvania, 1991.
74. Dolan, J. W., and Snyder, L. R., "Troubleshooting LC Systems." Humana Press, Clifton, New Jersey, 1989.

CHAPTER 4

Capillary Electrophoresis

4.1 Introduction

The electrophoretic separation technique is based on the principle that, under the influence of an applied potential field, different species in solution will migrate at different velocities from one another. When an external electric field is applied to a solution of charged species, each ion moves toward the electrode of opposite charge. The velocities of the migrating species depend not only on the electric field, but also on the shapes of the species and their environmment. Historically, electrophoresis has been performed on a support medium such as a semisolid slab gel or in nongel support media such as paper or cellulose acetate. The support media provide the physical support and mechanical stability for the fluidic buffer system. Capillary electrophoresis (CE) has emerged as an alternative form of electrophoresis, where the capillary wall provides the mechanical stability for the carrier electrolyte. Capillary electrophoresis is the collective term which incorporates all of the electrophoretic modes that are performed within a capillary.

4.2 Classification of Electrophoretic Modes

The most common classification scheme in electrophoresis focuses on the nature of electrolyte system. Using this scheme, electrophoretic modes are classified as continuous or discontinuous systems. Within these groupings the methods may be further divided on the basis of constancy of the electrolyte; if the composition of the background electrolyte is constant as in capillary zone electrophoresis, the result is a kinetic process. If the composition of the electrolyte is not constant, as in isoelectric focusing, the result is a steady-state process.

Alternatively, CE methods are classified according to the contribution of the electroosmotic flow to the separation process. For example, in the molecular sieving mechanism found in gel-filled capillaries, no osmotic flow is operative. In counterelectroosmotic CE, on the other hand, the electroosmotic flow flows in the opposite direction from the direction of the analyte ions. This classification is found in capillary ion electrophoresis (CIE), a technique which focuses on the separation of inorganic and low molecular weight organic species.

4.2.1 Classification According to Electrolyte System

A continuous electrolyte system is one in which the background electrolyte provides an electrically conducting and buffering medium along the migration path.[1] If the composition of the background electrolyte is constant, the electric field and the effective mobilities of the sample components will remain constant, and the sample components will migrate with mutually different but constant velocities. The result is a kinetic process, as exemplified by zone electrophoresis. If the composition of the background electrolyte is not constant, both the electric field and the effective mobilities of the sample components will vary along the migration path. As a consequence, the migration of the analytes through the electrolyte will be virtually nonexistent at some points along the separation path, resulting in different analytes with similar velocities. An example of a steady-state system is isoelectric focusing. The classification scheme is illustrated in Figure 4.1.

A discontinuous system is one in which the sample migrates as a distinct zone between two different electrolytes.[1] Unlike the continuous system, where the electrolyte is primarily responsible for the conduction of current in the sample zone, conduction of current through the sample in the discontinuous system is provided exclusively by the ions in the sample and the counterionic system. Isotachophoresis is an example of a discontinuous electrolyte system.

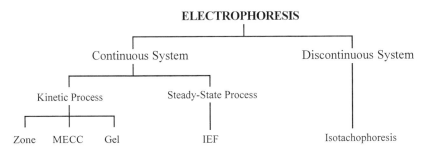

Figure 4.1 Classification of electrophoretic modes according to the nature of the electrolyte system. MECC, Micellar electrokinetic capillary chromatography; IEF, isoelectric focusing.

4.2.2 Classification According to Contribution of Electroosmotic Flow

In electrophoresis there are two main contributions to the movement of analyte ions: first, the electrophoretic mobility of the analyte ion itself and, second, the speed and direction of the electroosmotic flow (EOF). The EOF may be induced to travel in the same direction as the analyte ions or in the opposite direction from the analyte ions, or it can be suppressed so that the flow is negligible. Figure 4.2 shows the classification of electrophoresis according to the contribution of the electroosmotic flow.

If both the analytes and the EOF move in the same direction, as with CIE, the system is classified as coelectroosmosis, and the analytes reach the detector faster than they would as a result of their own mobilities. If the analytes move in the opposite direction from the analytes, the system is classified as counterelectroosmosis, and the analytes reach the detector later than they would as a result of their own mobilities. If the EOF is suppressed by eliminating the effective charge on the capillary wall, then the analytes reach the detector solely as a result of their own mobility. The last approach is often taken for the analysis of large peptides and proteins, where ionic or hydrophobic interactions between the analyte and the capillary wall result in peak tailing or total adsorption.

4.3 Basic Concepts of Capillary Electrophoresis

Capillary electrophoresis is similar to chromatography in many respects, and most of the words used in chromatography are also found in CE. For example, resolution and efficiency are common to both techniques and are defined in a similar way. However, some of the terminology is different, as illustrated in Table 4.1. For example, in chromatography, a column in used to separate the analytes; in electrophoresis, a capillary is used. In chromatography a pump is used to propel the sample through the column; in electrophoresis, there is no external pumping system, and the

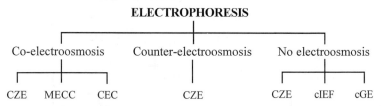

Figure 4.2 Classification of electrophoresis according to the contribution of the electroosmotic flow. CZE, Capillary zone electrophoresis; MECC, micellar electrokinetic capillary chromatography; CEC, capillary electrochromatography; cIEF capillary isoelectric focusing; cGE, capillary gel electrophoresis.

4.3 Basic Concepts of Capillary Electrophoresis

Table 4.1 Comparison of Electrophoretic and Chromatographic Terms

Capillary electrophoresis	Chromatography
Electropherogram	Chromatogram
Applied potential	Flow rate
Carrier electrolyte or buffer	Eluent or mobile phase
Injection mode (hydrostatic or electromigration)	Injector
Migration time	Retention time
Electrophoretic mobility	Column capacity factor
Velocity	—
Electroosmotic flow	—
High-voltage power supply	Pump
Capillary	Column

sample constituents move as a result of their mobilities in an applied potential field and the electroosmotic flow, if present.

4.3.1 Electrophoretic Mobility

The movement (migration) of a charged species under the influence of an applied field is characterized by its electrophoretic mobility, μ_e, which has units of $cm^2\ sec^{-1}\ V^{-1}$. Mobility is dependent not only on the charge density of the solute (the overall valence and size of the solute molecule), but also on the dielectric constant and viscosity of the electrolyte. Mobility is also strongly dependent on temperature, increasing by approximately 2% for each Kelvin rise in temperature.[2] In the presence of electroosmotic flow (Section 4.3.3), the "apparent" mobility is the sum of the electrophoretic mobility of the analyte, μ_e, and the mobility of the electroosmotic flow, μ_{eo}:

$$\mu = \mu_e + \mu_{eo}. \tag{4.1}$$

The apparent mobility, μ, may be determined, experimentally, using the equation[3]

$$\mu = l/tE = lL/tV \tag{4.2}$$

where l is the distance from the point of injection to the detector, t is the time taken for the species to migrate to the region of the detector, E is the electric field (V/L), V is the applied voltage and L is the distance between the two electrodes, that is, the total length of the capillary (if $l = L$, then the numerator of the right-hand term may be written as L^2). Thus, if the mobility of the electroosmotic flow is known, the electrophoretic mobility may be extracted. In the absence of any electroosmotic flow, the apparent mobility of the analyte, μ_e, will be equal to the electrophoretic mobility, μ.

4.3.2 Velocity

In electrophoresis, separation is based on differences in solute velocity in an applied electric field. The migration velocity of a particular species is the product of its mobility and the electric field. The migration velocity, v, may be written[4]

$$v = \mu_e E = \mu_e V/L. \quad (4.3)$$

In the presence of electroosmotic flow, Eq. (4.3) is more accurately written as

$$v = \mu E = \mu V/L. \quad (4.4)$$

4.3.3 Electroosmotic Flow

Electroosmotic flow (EOF) is the term used to describe the movement of a liquid in contact with a solid surface when a tangential electric field is applied.[5] This movement is also known as electroosmosis or electroendoosmosis. Electroosmotic flow can be eliminated if necessary. However, EOF is often used, when flowing in the same direction as the analytes, to increase the speed with which the analytes reach the detector or, when flowing in the opposite direction from the analytes, to improve resolution.

Electroosmosis occurs in fused silica capillaries because acidic silanol groups at the surface of the capillary dissociate when in contact with an electrolyte solution[6,7] (buffer), according to

$$SiOH(s) \rightleftharpoons SiO^-(s) + H^+(aq). \quad (4.5)$$

Hydrated cations in the electrolyte solution are attracted to the negatively charged silanol groups and become arranged into two layers. As illustrated in Figure 4.3, one layer is tightly bound by electrostatic forces (compact

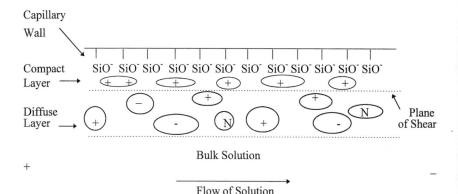

Figure 4.3 Electrical double layer at the capillary wall and creation of electroosmotic flow. (N = neutral analyte.)

layer), and the other is more loosely bound (diffuse layer). When an electric field is applied, the diffuse layer breaks away (at the plane of shear) and moves toward the cathode, dragging with it the bulk solution of the electrolyte, as a result of viscous drag. This flow of bulk solution is known as electroosmosis.

(i) Measurement of Electroosmotic Flow

To calculate the electrophoretic mobility of an analyte, the contribution from the electroosmotic flow to the apparent mobility must be known. The most common way to estimate the electroosmotic flow rate is to record the migration time of an injected uncharged marker solute,[8,9] which will be carried to the detector under the influence of only the EOF. The electroosmotic flow rate is obtained from the migration time of the neutral marker using Eq. (4.6)[10]:

$$\mu_{eo} = 1/Et. \tag{4.6}$$

Examples of neutral markers include water (in CIE), methanol, acetone, benzene, and dimethyl sulfoxide.

The rate of the EOF can also be determined by measuring the change in weight of the receiving electrolyte vial after a given time.[11,12] Using this technique, problems such as adsorption of the neutral marker are avoided. However, losses caused by evaporation must be eliminated, and the use of a digital balance is recommended. Alternatively, the EOF can be measured by the current monitoring method.[13] With this approach, the capillary and the receiving vial are filled with a buffer at concentration C. The injection vial is filled with a buffer at concentration $0.95C$. When a voltage is applied, the weaker buffer is pulled into the capillary by the EOF. Because the buffer is more dilute, the current will fall as buffer fills the capillary. The time taken for the current to stabilize, Δt, is measured and represents the time required to complete the filling of the capillary tube by the electroosmotic flow of the electrolyte. Thus, given knowledge of the total capillary length, L, between reservoirs 1 and 2, the electroosmotic flowrate is given by[13]

$$v_{eo} = L/\Delta t. \tag{4.7}$$

(ii) Factors Affecting Electroosmotic Flow

The speed of the EOF is directly related to the magnitude of the "zeta potential," ζ, which is given by[14]

$$\zeta = 4\pi\eta\mu_{eo}/\varepsilon \tag{4.8}$$

where η is the viscosity, ε is the dielectric constant of the buffer, and μ_{eo} is the electroosmotic flow mobility. Thus, factors affecting ζ will affect the EOF. The zeta potential is the potential at the shear plane (see Fig. 4.3) and is dependent on both the nature of the solid surface and the ionic state of the liquid. Polar solvents, such as water, give rise to zeta potentials as high as 100 mV when in contact with either polar surfaces such as glass or nonpolar surfaces such as graphite.[15] On the other hand, nonpolar, nonconducting solvents, such as heptane, do not normally exhibit a zeta potential.

A. *Effect of pH* Acidic silanol groups at the surface of the capillary wall will dissociate when in contact with an electrolyte solution, as illustrated by Eq. (4.5). At high pH, the silanol groups are fully ionized, generating a dense compact layer and a high zeta potential. As a result, the magnitude of the EOF in untreated fused silica capillaries increases with increasing pH.

B. *Effect of Ionic Strength* The dependence of mobility on ionic strength or concentration is illustrated by[16]

$$\mu = \frac{e}{(3 \times 10^7)|Z|\eta C^{1/2}} \qquad (4.9)$$

where Z is the number of valence electrons, e is the total excess charge in solution per unit area, and C is the buffer concentration. As shown in Figure 4.4, when the buffer concentration increases, the mobility of the EOF will decrease as the square root of the buffer concentration.

C. *Effect of Organic Solvent* The addition of organic solvents to buffers alters the viscosity of the buffer and may also affect the intramolecular hydrogen bonding. Therefore, organic modifiers such as acetonitrile and methanol are often added to buffers to improve resolution.

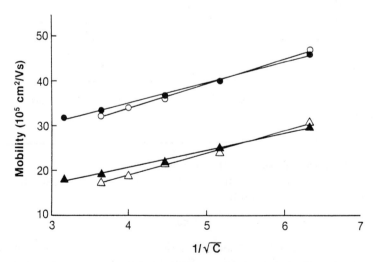

Figure 4.4 Mobility as a function of buffer concentration. Symbols: ○, μ_{eo} (acetate); ●, μ_{eo} (phosphate); △, μ_{ef} (dansylalanine, acetate); ▲, μ_{ef} (dansylalanine, phosphate). Capillary: 50 cm × 75 μm I.D. fused silica. (Reprinted from Ref. 16 with permission.)

4.3 Basic Concepts of Capillary Electrophoresis

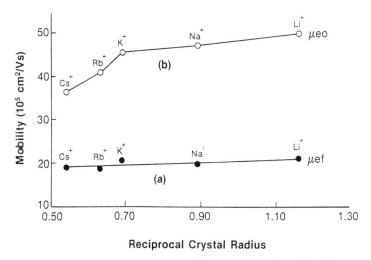

Figure 4.5 Effect of buffer cation on mobility. (a) Electrophoretic mobility (dansylalanine) versus reciprocal crystal radius; (b) electroosmotic mobility (mesityl oxide) versus reciprocal crystal radius. Capillary: 50 cm × 75 μm I.D. fused silica. (Reprinted from Ref. 16 with permission.)

D. Effect of Buffer Cation and Buffer Anion The electroosmotic flow is proportional to the potential drop across the diffuse layer of counterions associated with the capillary wall. Because the potential drop is formed by counterions in the buffer attracted to the charged silica surface, the nature of the counterions will affect the zeta potential and therefore the EOF. Figure 4.5 shows the effect of the buffer cation on the mobility of both the EOF (using mesityl oxide as the marker) and a solute (dansylalanine).[16] The highest mobility is obtained with the smallest cations; however, high mobility may decrease solute resolution, so care must be taken in choosing the cation. The buffer anion also affects the mobility of the EOF, although trends are less apparent. Therefore, the effect of the EOF on a separation can be altered by careful selection of both the buffer anion and cation.

E. Applied Voltage The mobility of the EOF increases with increasing applied voltage. This effect is attributed primarily to an increase in temperature with increasing applied voltage as a result of Joule heating. Joule heating is discussed in detail in section 4.3.4,i,D.

(iii) Manipulation of Electroosmotic Flow

Traditionally, CE has been applied to the analysis of comparatively large and multiply charged biomolecules using a standard instrument configuration (Chapter 6) in which the EOF flows toward the detector. As all the cationic species in solution will be carried toward the detector at an increased velocity as a result of the EOF, the time before they reach the detector will be reduced, and resolution will deteriorate. For that reason, and because of potential interactions of biomolecules with the capillary walls, much research has been directed toward alteration of the properties of capillary walls.

There are several ways to reduce or suppress the electroosmotic flow in capillaries. These methods involve either eliminating the zeta potential across the solution-solid interface or increasing the viscosity at this interface. One approach is to coat the capillary wall, physically, with a polymer such as methylcellulose or linear polyacrylamide. Because of the difficulty in deactivating the capillary surface reproducibly, however, alternative methods employing dynamic reduction of solute–capillary interactions have been developed. Dynamic reduction of these interactions include the addition of chemical reagents such as methylhydroxyethylcellulose, S-benzylthiouronium chloride, and Triton X-100.

Chemical derivatization of the capillary wall has also been used to alter the properties of the interface. Derivatization using trichlorovinylsilane followed by copolymerization with polyacrylamide is one approach. Another effective reagent is (α-methacryloxypropyl)trimethoxysilane, followed by cross-linking the surface-bound methacryl groups with polyacrylamide.

In some instances, such as in the separation of inorganic ions by CIE, complete elimination of the EOF is undesirable. Changing the velocity of the EOF or reversing the direction may be preferable. For cation analyses, a standard instrument configuration is used with a positive power supply. Under these conditions, the EOF naturally travels toward the detector, in the same direction as the cations. For anion analysis, the instrument configuration is reversed, so that the power supply is negatively charged. To take advantage of the EOF, the charge on the capillary must be reversed by the addition to the electrolyte solution of cationic surfactants such as cetyltrimethylammonium bromide (CTAB) and tetradecyltrimethylammonium bromide (TTAB). The surfactants coat the inner capillary wall, as illustrated in Figure 4.6, resulting in a capillary wall with a positive charge instead of the negative charge seen with uncoated fused silica. On application of a potential gradient, the predominantly positively charged double layer migrates towards the anode, in the same direction as the analytes.

Direct control of the EOF in capillary zone electrophoresis can be obtained by using an external electric field. The EOF may be increased, decreased, or even reversed in the fused silica capillaries by the application of a separate potential field across the wall of the capillary. Further, the zeta potential can be changed at any time during the analysis to achieve innovative separation results.

4.3 Basic Concepts of Capillary Electrophoresis

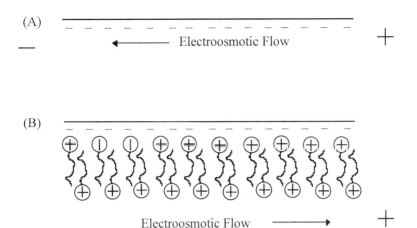

Figure 4.6 Schematic illustration of the capillary wall with (A) no surfactant added and (B) excess surfactant added to reverse the EOF.

4.3.4 Separation Efficiency

Because the electroosmotic flow affects the amount of time a solute resides in the capillary, both the separation efficiency and resolution are related to the direction and flow of the EOF. The EOF flow profile, as shown in Figure 4.7, is comparatively pluglike. Unlike the laminar flow that is characteristic of pressure-driven fluids,[5] the EOF has minimal effect on resistance to mass transfer. As a result, the plate count in a capillary is far larger than that of a chromatography column of comparable length.

The migration time of a solute, or the time t required for a zone to migrate from the point of injection to the point of detection, is given by

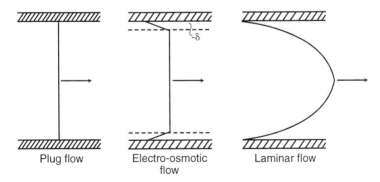

Figure 4.7 Flow profile under various types of flow.

$$t = l/v = l/\mu E = lL/\mu V. \tag{4.10}$$

Diffusion will occur during that time. According to Einstein's equation, if an infinitesimal band is allowed to diffuse for a time t, the spatial variance, σ_L^2, of the zone will be given by

$$\sigma_L^2 = 2Dt = 2DL^2/\mu V \tag{4.11}$$

where D is the diffusion coefficient of the solute. The separation efficiency of an electrophoretic system may be expressed in terms of the number of theoretical plates, N, where

$$N = L^2/\sigma_L^2 \tag{4.12}$$

and, by substituting Eq. (4.11) into Eq. (4.12),

$$N = \mu V/2D. \tag{4.13}$$

Three points can be drawn from Eq. (4.13): first, the most direct approach to high separation efficiencies in zone electrophoresis is the use of very high voltages (V); second, large molecules, such as DNA and proteins, which have low diffusion coefficients (D), will give high efficiencies because they exhibit less dispersion than small molecules; and, third, highly mobile species (μ) will produce high plate counts because the rapid velocity through the capillary minimizes the time for diffusion. Although the last two points appear contradictory, Eq. (4.13) illustrates the wide range of molecular weights across which high-efficiency separations are possible in CE. As long as heat dissipation is adequate, capillary length plays no direct role in separation efficiency.

The efficiency may be determined experimentally using

$$N = 5.54 \, (t/w_{0.5})^2 \tag{4.14}$$

where t is the migration time and $w_{0.5}$ is the width of the peak at half height. Equation (4.14) is strictly valid only for Gaussian peaks, and any peak asymmetry should be taken into account, for example, by the use of central moments.

(i) Factors Affecting Separation Efficiency

In the discussion of separation efficiency thus far, it has been assumed that one-dimensional diffusion is the major contributor to peak broadening in CE, and other factors that contribute to the diffusion process have been ignored. In reality, the total variance of a zone, once corrections have been made for zone velocity and finite detector width, is given by

$$\sigma^2 = \sigma_i^2 + \sigma_d^2 + \sigma_e^2 + \sigma_t^2 + \sigma_{eo}^2 + \sigma_o^2 \tag{4.15}$$

where the right-hand terms represent the contributions of injection, diffusion, electromigration, temperature profiles due to Joule heating, electroosmosis, and other effects such as interactions between the analytes and the capillary wall, respectively. Except for σ_i^2, the variances are directly proportional to the analysis time.

The first four terms on the right-hand side of Eq. (4.15) represent effects inherent in the principle of the method and cannot be suppressed to zero; however, their influence on the separation efficiency can be controlled by instrument design and selection of appropriate working conditions. Interactions between the analyte and capillary wall is a concern primarily for the analysis of biomolecules, and they may be reduced or eliminated by coating the capillary wall.

A. Injection Length The extent to which diffusion affects the zone length depends on the volume of the sample plug that is injected, and in particular on the length of the injected plug. For small injection lengths, the zone length is determined almost exclusively by diffusion, and it is independent of injection length. For large injection lengths, diffusional broadening may be neglected, and the zone length becomes equal to twice the square root of the variance of the injection length. At intermediate injection length, both diffusion and injection length contribute to the zone length.

In practice, the width of the initial sample pulse should be less than 1% of the total length of the capillary. However, at very small injection lengths, the amount actually injected is dominated by mixing at the interface between the capillary and the sample solution. The minimum injection length for a capillary with 75 μm I.D. has been estimated at approximately 200 μm,[17] although typical injection lengths are 5–10 times longer.

B. Diffusion There are three basic sources of intracolumn band broadening, as discussed in Chapter 2: eddy diffusion, longitudinal diffusion, and resistance to mass transfer. The first source, eddy diffusion, is characterstic of flow in a packed chromatographic bed and does not exist in hollow, open tubular capillaries. The third source, resistance to mass transfer, is also of little concern in electrophoresis, since there is no column packing to interfere with migration of the solutes toward the detector. Thus, in electrophoresis, longitudinal diffusion is the only source of diffusion that plays a significant role in band broadening.

Longitudinal diffusion arises from the random Brownian motion that all solutes undergo in all directions as they travel along a column. Longitudinal diffusion (along the axis of the column) results in longer bands and hence poorer resolution. As indicated in the discussion of injection length, diffusion is only limiting for very small injection lengths ($\geq 2\%$ of the total capillary length).

C. Electromigration Dispersion Electromigration dispersion manifests itself in the form of either fronting or tailing peaks, as shown in Figure 4.8. The peak shapes occur as a result of conductivity differences between the analyte zones and the carrier electrolyte (buffer). Conductivity differences

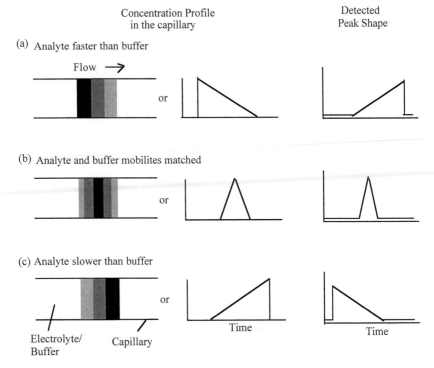

Figure 4.8 Electromigration dispersion caused by mismatched analyte and electrolyte mobilites.

can be present, either as a result of differences between the mobilities of the analyte zones with respect to the carrier electrolyte or as a result of concentration differences between the analyte zones and the carrier electrolyte. Two theories have been proposed to explain the peak shapes. The first, by Mikkers et al.,[18] addresses the mobility differences, and the second, by Hjerten,[19] addresses the concentration differences.

The peak shape distortions in CE can be explained, most simply, in terms of conventional circuit theory, where

$$V = IR \quad (4.16)$$

and I represents current, V is voltage, and R is resistance. The current along the length of the capillary remains constant. When the capillary is filled only with the carrier electrolyte, the voltage along the capillary can also be assumed to be constant. However, when a sample containing solutes of different mobilities is introduced into the capillary, the solutes will separate into zones, dependent on their relative mobilities. Each solute zone will be separated from the next by a zone of carrier electrolyte, and the voltage along the capillary will no longer be constant.

In zones of high conductivity or low resistance (high mobility species), the voltage will be decreased. The solutes will diffuse forward into the electrolyte zone where they will experience a higher voltage owing to the lower conductivity of the electrolyte. The solute ions will therefore continue to move forward, and the leading edge of the solute zone will become diffuse, leading to fronting as shown in Figure 4.8a. When the tailing edge of the solute zone encounters the higher voltage, it accelerates forward into the solute zone, maintaining a sharp tailing edge.

In zones of low conductivity or high resistance (low mobility species) the voltage will be increased. Although the solutes will diffuse forward into the electrolyte zone, in moving forward they encounter a region of lower voltage owing to the higher conductivity of the electrolyte. As a result of the lower voltage, the solutes slow down and reenter their own solute zones, thereby maintaining a sharp leading edge. Meanwhile, the electrolyte zones following behind each of the slower solute zones diffuse forward into the solute zones. Owing to the increased voltage experienced by the electrolyte in the slower solute zones, the electrolyte coions accelerate further into the solute zones, resulting in diffuse tailing edges to the slower solute zones (Fig. 4.8c). Solute zones that have a mobility similar to that of the carrier electrolyte are unaffected by conductivity differences, and therefore appear as symmetrical peaks (Fig. 4.8b).

Electromigration dispersion (electrodispersion), therefore, is a diffusion process that occurs because of changes in the local field strength in the migrating zone with respect to that of the background electrolyte.[20] It can be minimized by decreasing the difference between the mobility of the solute and that of the background electrolyte coion, or by decreasing the concentration of the separated zones.[21] Usually, the electromigration dispersion is negligible when the concentration of the analyte is more than two orders of magnitude lower than that of the background electrolyte.[20] This condition can be fulfilled in two ways: by injecting a small amount of a dilute sample or by using a high concentration background electrolyte. Both approaches, however, have limitations. When weakly absorbing species are detected using a UV absorbance detector, the sample concentration cannot be reduced below the sensitivity of the detector used. At the same time, a substantial increase in the concentration of the background electrolyte (BGE) will cause an increase in the BGE conductivity, which will cause other problems.

D. Joule Heating From Einstein's equation for longitudinal diffusion [Eq. (4.11)], a simple expression for N was obtained [Eq. 4.13)] which suggests that high voltages should be applied, in order to minimize diffusion. However, high power inductions cause excessive band broadening because of Joule heating of the liquid inside the capillary. Thus, a compromise is required between band broadening at lower voltages caused by diffusion and band broadening at higher voltages caused by Joule heating.

Joule or ohmic heating is the name given to the heat generation caused by collisions between solute ions and electrolyte ions as a result of the conduction of electric currents. It can cause nonuniform temperature gradients and local changes in viscosity, both of which lead to zone broadening. Without adequate heat dissipation, the heat generated by the use of high separation voltages can rapidly evaporate any electrolyte solution. The narrower the capillary inner diameter, the more efficient is the heat dissipation.

Because the capillary is cooled only on the outer surface, the temperature difference between the solution in the center and the inner capillary wall can be significant, even in very narrow capillaries. As a consequence, the viscosity of the electrolyte will be lower at the center of the capillary, about 2.7% per degree Celsius, and the migration velocity of the analytes will be higher.[19] Lukacs et al.[22] investigated the relationship between efficiency and capillary diameter and also that between efficiency and capillary length, at constant voltage (15 kV), in Pyrex capillaries. They noted that the narrower the capillary, the more efficient was the separation, down to diameters of approximately 80 μm; below 80 μm, the increase in efficiency was more gradual, suggesting that a minimum temperature gradient is reached. Lukacs et al.[22] also reported an increase in efficiency when the capillary length is increased, to a maximum length L of 90 cm. At this "critical" length, they assumed that the electric field was low enough that the capillary could provide adequate cooling, and that thermally induced broadening leveled off.

Although thermal effects can be reduced by the use of narrow-bore capillaries with low conductivity or low concentration electrolytes, those approaches have other consequences. Electrolytes of low concentration limit sample loading, whereas decreased capillary radius increases the capillary surface area-to-volume ratio, which can enhance the potential for adsorption effects.[23] In addition, the concentration sensitivity and signal-to-noise ratio for optical detectors will decrease with decreasing column diameter.

A simple, practical approach to determine the optimum voltage without exceeding the heat removal capacity of the system is to draw an Ohm's law plot, as illustrated in Figure 4.9.[24] The capillary is filled with the carrier electrolyte, voltage is applied, and the resulting current is recorded. The voltage is then varied and the current recorded at each new voltage. When the current is plotted against the applied voltage, a straight line should result. A positive deviation from linearity shows that the temperature removal capacity of the system has been exceeded.

E. Solute–Wall Interactions Interaction between the solute and the capillary wall leads to band broadening, peak tailing, and irreproducibility

4.3 Basic Concepts of Capillary Electrophoresis

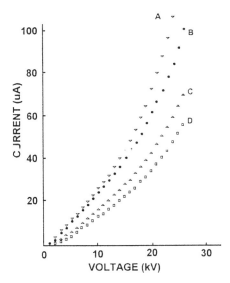

Figure 4.9 Ohm's law plots showing the optimum voltages for different capillary temperatures using air circulation for temperature control. (Reprinted from Ref. 24 with permission.) (A) No control; (B) 25°C; (C) 10°C; (D) 4°C.

of separations. The primary causes of adsorption are hydrophobic interactions and ionic interactions between some cationic solutes, such as proteins and the negatively charged capillary wall. At most pH values, the wall of the capillary is negatively charged because of silanol ionization. At pH values below the pI, a protein exists in a cationic form that will ion-pair to the negatively charged capillary wall. Several strategies have been proposed to minimize solute–wall interactions.

Probably the most effective way to reduce solute–wall interactions is to coat the capillary wall. Coatings can take several forms including dynamically coated capillaries, where the coating is added to the electrolyte, and functionalized capillaries, where the coating is covalently bonded to the capillary. Another approach is to increase the ionic strength of the buffer. Increased ionic strength has been found to decrease adsorption effects, owing to competition for potential adsorption sites. However, increasing the ionic strength can be problematic because of Joule heating, which dictates lower applied potentials, so a compromise is required.

A third option for reducing solute–wall interactions is to operate at extremes of pH. At pH 1.5, the silanol groups at the capillary wall are not ionized. Although proteins are cationic at that pH, electrostatic interactions are eliminated. Operating at pH values that are within 1–2 units of the pI of the protein will also reduce wall effects, but these approaches are limited by hydrophobic interactions and poor selectivity.

4.3.5 Resolution

Resolution in CE is a function of three parameters: selectivity, column efficiency, and migration time.[25] Each of the parameters is influenced by many factors, including applied voltage, electrolyte pH, ionic strength, and treatment of the inner capillary wall. The pH is the dominant factor that can be manipulated to control resolution.

Resolution, R_s, is defined as

$$R_s = \frac{2(t_2 - t_1)}{w_1 + w_2} = \frac{t_2 - t_1}{4\sigma} \qquad (4.17)$$

where t is the migration time, w is the baseline width in time, and σ is the temporal standard deviation. The resolution between two solutes may also be defined in terms of efficiency according to

$$R_s = \frac{l(\mu_2 - \mu_1)N^{1/2}}{4(\overline{\mu} + \mu_{eo})}. \qquad (4.18)$$

By substituting Eq. (4.18) and the plate count equation [Eq. (4.13)] into the expression for resolution, the resolution equation may be rearranged to yield an expression that better describes the effect of various parameters on resolution:

$$R_s = 0.177(\mu_2 - \mu_1)\left[\frac{V}{D(\overline{\mu} + \mu_{eo})}\right]^{1/2}. \qquad (4.19)$$

It can be seen from Eq. (4.19), that increasing the voltage is not a very efficient means of increasing the resolution; because of the square root relationship, the voltage must be quadrupled to double the resolution. Since the voltage is usually in the range 10 − 30 kV, Joule heating limitations are rapidly approached. Equation (4.19) also suggests that decreasing the contribution from the EOF, or adjusting the EOF so that it flows in the opposite direction from the electrophoretic flow, will increase resolution. Under these conditions, the "effective" length of the capillary is increased, thereby improving resolution. However, the analysis time will be increased, so it is important to balance resolution and analysis time.

4.4 Summary of Major Concepts

1. Capillary electrophoresis methods may be classified according to the nature of the electrolyte system or according to the contribution of the electroosmotic flow. The former classification is the most popular, and two types of systems have been identified:

continuous systems (including kinetic and steady-state processes) and discontinuous systems.

4.4 Summary of Major Concepts

Those systems form the basis of the electrophoretic modes discussed in Chapter 5.

2. Electrophoresis is described and measured in terms of five major concepts:

mobility,
velocity,
electroosmotic flow,
efficiency, and
resolution.

The electrophoretic mobility of a solute is a characteristic property of the solute that describes its movement under the influence of an applied field. Electrophoretic mobility is dependent both on the charge density of the solute and on physical properties of the buffer system. The apparent mobility of a species, that is, the mobility that the species appears to possess, is the sum of the electrophoretic mobility and the mobility of the electroosmotic flow.

The migration velocity of a species is the product of its electrophoretic mobility and the electric field. In the presence of electroosmotic flow, the velocity is the product of the apparent mobility and the electric field.

Electroosmotic flow (EOF, electroosmosis, electroendoosmosis) is the term used to describe the movement of bulk solution in contact with a solid surface, when a tangential electric field is applied. Electroosmotic flow is used to improve separations of inorganic species in capillary ion electrophoresis, but it is often suppressed during the analysis of larger biomolecules in order to improve resolution.

Efficiency, N, is a number that describes peak broadening as a function of migration time. High voltages, fast migration times, and species with low diffusion coefficients result in the most efficient separations.

Resolution, R, is dependent on both the efficiency of the system and the differences in the migration times of the analytes of interest. Resolution is affected by the speed and direction of the EOF, the mobility of the analytes of interest, and, to a lesser degree, the applied voltage.

3. The velocity of the electroosmotic flow can be affected by altering the properties of the buffer system or by changing the applied voltage. The factors involved are

pH (EOF increases with increasing pH),
ionic strength (as ionic strength increases, EOF decreases as the square root of the ionic strength),
organic solvent (alters buffer viscosity and affects hydrogen bonding),
buffer cation and anion, and
applied voltage (increases with increasing voltage)

4. The efficiency of the system is affected by

injection length,
diffusion,
electromigration dispersion,
Joule heating, and
solute-wall interactions.

For small injection lengths the peak widths are determined by diffusion. For large injection lengths, peak widths are determined by two times the square root of the variance of the injection length.

Longitudinal diffusion is the major source of diffusion. Eddy diffusion and resistance to mass transfer contribute only to packed columns.

Electromigration dispersion determines the shapes of the peaks. Unless the mobilites of the analytes match that of the electrolyte, or unless the concentration of the analytes is at least two orders or magnitude lower than that of the buffer, the peaks will front or tail; slower analytes tail while faster analytes front.

Joule heating is the increase in temperature that results when analyte and buffer molecules collide as a result of the conduction of electric currents. It does not adversely affect efficiency if the capillary length is less than 90 cm and the internal diameter is less than 80 μm.

The most effective way to eliminate solute–wall interactions is to coat the capillary wall and thereby eliminate charged attraction sites. Increasing the ionic strength of the buffer or operating at the pH extremes can also help.

References

1. Kleparnik, K., and Bocek, P., *J. Chromatogr.* **569**, 3 (1991).
2. Wieme, R. J., in "Chromatography: A Laboratory Handbook of Chromatography and Electrophoresis Methods" (E. Heftmann, ed.), 3rd Ed. Van Nostrand, New York, 3rd ed., 1975.
3. Cohen, A. S., Paulus, A., and Karger, B. L., *Chromatographia* **24**, 15 (1987).
4. Jorgenson, J. W., and Lucaks, K. D., *Anal. Chem.* **53**, 1298 (1981).
5. Pretorius, V., Hopkins, B. J., and Schieke, J. D., *J. Chromatogr.* **99**, 23 (1974).
6. Schwer, C., and Kenndler, E., *Chromatographia* **30**, 546 (1990).
7. Hayes, M. A., and Ewing, A. G., *Anal. Chem.* **64**, 512 (1992).
8. Lauer, H. H., and McManigill, D., *Anal. Chem.* **58**, 165 (1986).
9. Walbroehl, Y., and Jorgenson, J. W., *Anal. Chem.* **58**, 479 (1986).
10. Hanai, T., Hatano, H., Nimura, N., and Kinoshita, T., *HRC & CC* **14**, 481 (1991).
11. Altria, K. D., and Simpson, C. F., *Chromatographia* **24**, 527 (1987).
12. Altria, K., and Simpson, C., *Anal. Proc.* **23**, 453 (1986).
13. Huang, X., Gordon, M. J., and Zare, R. N., *Anal. Chem.* **60**, 1837 (1988).
14. Adamson, A. W., "Physical Chemistry of Surfaces," 2nd Ed., Chapter 4. Wiley (Interscience), New York, 1967.

15. Davies, J. T., and Rideal, E. K., "Interfacial Phenomena," Chapter 3. Academic Press, New York, 1961.
16. Issaq, H., Atamna, I., Muschik, G., and Janini, G., *Chromatographia* **32**, 155 (1991).
17. Huang, X., Coleman, W. F., and Zare, R.,N., *J. Chromatogr.* **480**, 95 (1989).
18. Mikkers, F. E. P., Everaerts, F. M., and Verheggen, T. P. E. M., *J. Chromatogr.* **169**, 11 (1979).
19. Hjerten, S., *Electrophoresis* (*Weinheim*) **11**, 665 (1990).
20. Mikkers, F. E. P., Everaerts, F. M., and Verheggen, T. P. E. M., *J. Chromatogr.* **169**, 1 (1979).
21. Foret, F., Deml, M., and Bocek, P., *J. Chromatogr.* **452**, 601 (1988).
22. Lukacs, K. D., and Jorgenson, J. W., *HRC & CC* **8**, 407 (1985).
23. Jorgenson, J. W., and Lucaks, K. D., *Science* **222**, 266 (1983).
24. Kurosu, Y., Hibi, K., Sasaki, T., and Saito, M., *HRC & CC* **14**, 200 (1991).
25. Cohen, A. S., and Karger, B. L., *J. Chromatogr.* **397**, 409 (1987).

CHAPTER 5

Separations in Capillary Electrophoresis

5.1 Introduction

In capillary electrophoresis a sample, usually containing charged species, is introduced into the end of a capillary that has been filled with a solution of buffer (or electrolyte). Under the influence of an electric field, the analytes migrate away from the injection end of the capillary toward the detector end, where they are visualized. Three distinct separation mechanisms have been developed for the separation of analytes by CE.

The first and most often encountered separation mechanism in CE is based on mobility differences of the analytes in an electric field; these differences are dependent on the size and charge-to-mass ratio of the analyte ion. Analyte ions are separated into distinct zones when the mobility of one analyte differs sufficiently from the mobility of the next. This mechanism is exemplified by capillary zone electrophoresis (CZE) which is the simplest CE mode. A number of other recognized CE modes are variations of CZE. These are micellar electrokinetic capillary chromatography (MECC), capillary gel electrophoresis (CGE), capillary electrochromatography (CEC), and chiral CE. In MECC the separation is similar to CZE, but an additional mechanism is in effect that is based on differences in the partition coefficients of the solutes between the buffer and micelles present in the buffer. In CGE the additional mechanism is based on solute size, as the capillary is filled with a gel or a polymer network that inhibits the passage of larger molecules. In chiral CE the additional separation mechanism is based on chiral selectivity. Finally, in CEC the capillary is packed with a stationary phase that can retain solutes on basis of the same distribution equilibria found in chromatography.

The second separation mechanism is found in capillary isoelectric focusing (cIEF), where analytes are separated on the basis of isoelectric points. The third mechanism is found in capillary isotachophoresis (cITP), where all of the solutes travel at the same velocity through the capillary but are separated on basis of differences in their mobilites. This chapter describes the CE modes in terms of the separation mechanisms and the buffer systems. Examples of compounds separated by each mode, and the advantages and disadvantages of each mode, are provided to enable investigators to choose the most appropriate mode for various applications. By the end of this chapter readers should understand the factors affecting the separation and be able to develop basic separations in each mode.

5.2 Capillary Zone Electrophoresis

Capillary zone electrophoresis (CZE) is the most simple and widely used mode in CE. Separations take place in an open-tube, fused silica capillary under the influence of an electric field. The velocity of the analytes is modified by controlling the pH, viscosity, or concentration of the buffer, or by changing the separation voltage. The electroosmotic flow is often used in this mode to improve resolution or to shorten analysis times.

5.2.1 Mechanism of Separation

In CZE the composition of the electrolyte is constant in the capillary and the reservoirs surrounding the two electrodes. The electrolyte provides an electrically conducting and buffering medium (continuous system). On introduction of a sample, each species of analyte ions migrates in the buffer in a discrete zone and at a different velocity from the other species. Neutral molecules are carried along by the EOF as a single, unresolved peak. Owing to the presence of the EOF, both anions and cations can be seen in a single run; however, separations are usually optimized for one species or the other. The mechanism is illustrated in Figure 5.1.

5.2.2 Buffer Systems

Buffers are usually chosen on the basis of the required pH; some buffers, however, are particularly useful for specific applications. Phosphate,

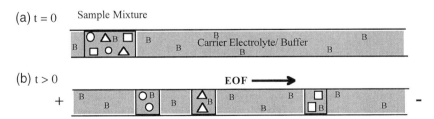

Figure 5.1 Separation mechanism in CZE, where B represents buffer. (a) Sample mixture is introduced into the capillary; (b) voltage is applied, and the analytes separate into distinct zones.

Table 5.1 Common Buffers for Capillary Zone Electrophoresis[a]

Name	pK_a	Mobility (cm^2/vs) × 10^{-4}
Phosphate	2.12 (pK_{a_1})	
Citrate	3.06 (pK_{a_1})	
Formate	3.75	−56.6[b]
Succinate	4.19 (pK_{a_1})	
Citrate	4.74 (pK_{a_2})	
Acetate	4.75	−42.4[b]
Citrate	5.40 (pK_{a_3})	
Succinate	5.57 (pK_{a_2})	
MES	6.15	−26.8[c]
ADA	6.60	
PIPES	6.80	
Imidazole	7.00	
Phosphate	7.21 (pK_{a_2})	
HEPES	7.55	−21.8[c]
HEPPS	8.00	
Tricine	8.15	
Tris	8.30	
Morpholine	8.49	
Borate	9.24	
CHES	9.50	
CHAPS	10.40	
Phosphate	12.32 (pK_{a_3})	

[a] Adapted from Ref. 4 with permission.
MES; 2-(N-morpholino)ethanesulfonic acid.
ADA; N-[2-acetamido]-2-iminodiacetic acid
PIPES; Piperazine-N,N'-bis-[ethanesulfonic acid]
HEPES; N-2-hydroxyethylpiperazine-N'-2-ethanesulfonic acid
HEPPS; N-[2-hydroxyethyl]piperazine-N'-[2-hydroxypropanesulfonic acid]
TRICINE; N-[2-(hydroxymethyl)ethyl]-glycine
TRIS; Tris(hydroxymethyl)amino methene
CHES; 2-[N-cyclohexylamino]ethanesulfonic acid
CHAPS; 3-[(3-cholamidopropyl)-dimethylammonio]-1-propanesulfonate
[b] *J. Chromatogr.* **390**, 69 (1987).
[c] *Chem Rev.* **89**, 419 (1989).

for example, is used for low-pH protein separations,[1] whereas borate is used for the separation of carbohydrates[2] and catecholamines.[3] It is important to have a well-buffered system with little or no absorbance at the detection wavelength. A list of commonly used buffers, with pK_a values and mobilities, is given in Table 5.1.[4]

Once the buffer has been chosen, the velocity of the analytes is modified

5.2.3 Practical Considerations

Mobility differences can be optimized for a given separation by controlling the operating voltage, electrolyte composition, pH, ionic strength, additives, and capillary wall coatings. The first step in methods development is to ensure adequate buffer ionic strength in order to minimize loading effects (Section 4.3.4.C). Figure 5.2 shows the effect of ionic strength on

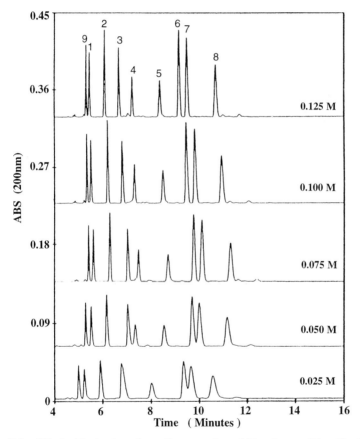

Figure 5.2 Effect of ionic strength on the separation of bioactive peptides. Conditions: capillary, 375 μm O.D. × 75 μm I.D. × 57 cm (50 cm to detector); separation voltage, 30 kV; temperature, 30°C; buffer, 0.025 M–0.125 M sodium dihydrogen phosphate, pH 2.44. (Reprinted from Ref. 5 with permission.) (1) Bradykinin (BRAD); (2) Angiotensin II(ANG II); (3) Thyrotropin releasing hormone (TRH); (4) Luteinizing hormone releasing hormone (LHRH); (5) Bombesin (BOMB); (6) Leucine enkephalin (LENK); (7) Methionine enkephalin (MENK); (8) Oxytocin (OXYT); (9) Dynorphin (DYNO).

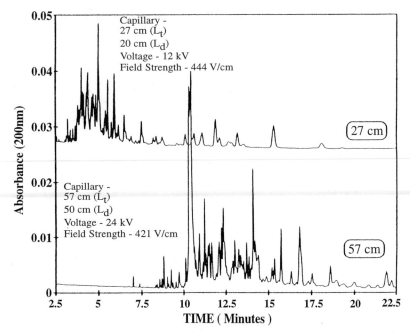

Figure 5.3 Effect of capillary length on the resolution of tryptic fragments of BSA. Conditions: capillary, 375 μm O.D. × 50 μm I.D.; buffer, 100 mM sodium dihydrogen phosphate, pH 2.5; temperature, 25°C; detection, UV absorbance at 200 nm; injection, pressure for 10 sec; sample, 10/1 dilution of BSA digest. (Reprinted from Ref. 5 with permission.)

the separation of bioactive peptides.[5] As the ionic strength is increased up to a limit, the number of theroretical plates increases, resulting in better resolution and improved peak shape. However, more heat is produced as the ionic strength is increased so effective temperature control is necessary; smaller diameter capillaries allow the use of higher ionic strengths.

If pH and ionic strength are insufficient to resolve all the analytes, other parameters can be changed. The length of the capillary has a large impact on resolution because the theoretical plate count increases with capillary length. Although resolution only increases with the square root of the capillary length, increasing the capillary length can be sufficient to produce the desired resolution. Figure 5.3 shows the effect of capillary length on the resolution of tryptic fragments of bovine serum albumin (BSA) at constant field strength, where the field strength is the voltage (volts) divided by the capillary length (centimeters).

Increasing the voltage has little effect on resolution but impacts the analysis time. The higher the separation voltage, the faster the migration time. However, care must be taken to prevent excessive voltage, which causes Joule heating, lowers the number of theroretical plates, and lowers

5.2 Capillary Zone Electrophoresis

Table 5.2 Buffer Additives for Capillary Zone Electrophoresis[a]

Additive	Function
Inorganic salts	Induce protein conformational changes, reduce wall interactions
Zwitterions	Reduce wall interactions
Crown ethers	Modify mobility
Organic solvents	Solubilize samples, modify EOF, reduce wall interactions
Urea	Solubilize proteins
Metal ions	Modify mobility
Sulfonic acids	Act as ion pairing agents, surface charge modifiers
Cellulose polymers	Tie up active sites on capillary wall
Cationic surfactants	Reverse charge on capillary wall
Organic acids	Modify mobility

[a] Adapted from Ref. 6 with permission.

the resolution. Finally, if the desired resolution has still not been obtained, additives can be introduced. Table 5.2. lists various buffer additives and their functions.[6]

Figure 5.4 shows the separation of histidine-containing dipeptides by CZE using a buffer containing zinc ions.[7] Copper ions or zinc ions in acetate or phosphate buffers (pH 2.5) were used, with good results. In the absence of metal ions, little resolution was evident, but as the metal ion concentra-

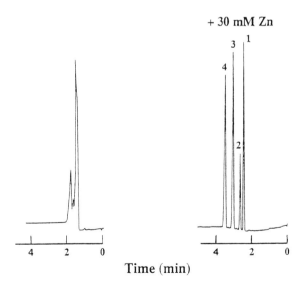

Figure 5.4 Electrophoretic behavior of (1) β-Ala-L-His, (2) L-Ala-L-His, (3) L-His-L-Ser, and (4) L-His-L-Glu in (a) the absence and (b) presence of 30 mM zinc sulfate, in a background electrolyte of 0.1 M acetic acid. (Reprinted from Ref. 7 with permission.)

tion was increased in 5 mM increments, resolution gradually improved. The buffer was also applied to the separation of some histidine-containing dipeptide isomers with good effect.

Capillary zone electrophoresis has been used for the separation of a wide variety of analytes, ranging from inorganic ions[8-10] to proteins,[11] peptides,[12,13] and nucleotides.[14] Protein and peptide separations are by far the most common application areas, and CE is able to provide information complementary to that obtained by HPLC.

Figure 5.5 shows the separation of recombinant extracellular superox-

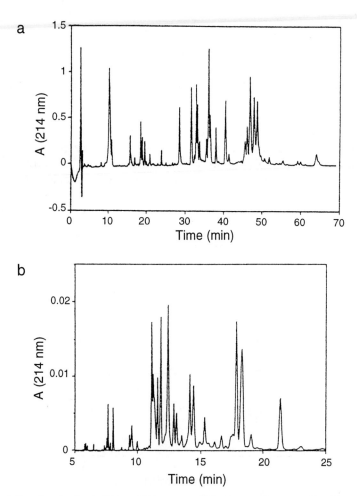

Figure 5.5 Tryptic digest of extracellular superoxide dismutase separated by (a) reversed-phase HPLC, using 0.1% (v/v) trifluoracetic acid in an acetonitrile/water gradient, and (b) CZE, using a 100 mM phosphate buffer, pH 2.5. (Reprinted from Ref. 15 with permission.)

ide dismutase using reversed-phase HPLC and CZE.[15] The protein was purified to more than 98% purity, carboxymethylated, and cleaved with trypsin prior to injection. The buffer for CZE was 100 mM phosphoric acid, pH 2.5, and as can be seen in Figure 5.5, a very different profile is provided by the CZE separation. Fractions from an SEC separation were purified by reversed-phase HPLC and used to identify some of the peptides in the CZE run.

5.3 Micellar Electrokinetic Capillary Chromatography

Micellar electrokinetic capillary chromatography (MECC) is a mode of CE similar to CZE, in which surfactants (micelles) are added to the buffer system. Micellar solutions can be used to solubilize hydrophobic compounds that would otherwise be insoluble in water. In MECC the micelles are used to provide a reversed-phase character to the separation mechanism. Although MECC was originally developed for the separation of neutral species by capillary electrophoresis, it has also been shown to enhance resolution in the analysis of a variety of charged species.[16]

5.3.1 Mechanism of Separation

The separation of neutral species is accomplished in MECC by the addition of surfactants into the carrier electrolyte. When an ionic surfactant is placed in the buffer at a concentration exceeding the critical micelle concentration (CMC), the surfactant monomers aggregate to form micelles, as illustrated in Figure 5.6. The micelles are essentially spheres in which the hydrophobic tails of the surfactants are oriented toward the center and the charged head groups face outward into the electrolyte solution. The separation mechanism is based on the differential partition of the solutes between the hydrophobic interior of a charged micelle and the aqueous phase. MECC is commmonly performed with anionic surfactants, the surfaces of which have a net negative charge. Sodium dodecyl sulfate (SDS) is the most commonly cited surfactant for use in MECC, owing to its high water solubility and lipid-solubilizing power.

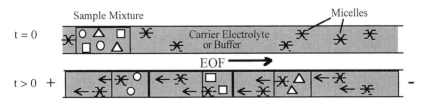

Figure 5.6 Schematic illustration of the separation mechanism in MECC.

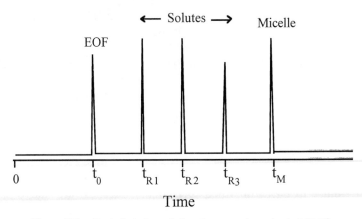

Figure 5.7 Typical elution window for neutral species in MECC.

When an anionic surfactant such as SDS is used, the micelles migrate toward the anode as a result of electrophoretic mobility. The EOF in uncoated, fused silica capillaries, however, travels toward the cathode at a greater velocity than the micelles travel toward the anode, thereby dragging the micelles slowly toward the detector. When a neutral molecule is injected into the system, it partitions itself between the buffer and the micelle. The more time it spends in the micelle, the longer it will take to reach the detector. In MECC, therefore, the analytes all migrate between a water peak, which is totally unretained, and the micelle peak, which is the most retained. A typical elution order is shown in Figure 5.7, where peak EOF represents a neutral molecule that has no interaction with the micelles and therefore travels at the velocity of the EOF.

For electrically neutral molecules eluting between t_0 and t_M (Fig. 5.7), the capacity factor, k', is given by[17]

$$k' = \frac{t_R - t_0}{t_0(1 - t_R/t_M)} \qquad (5.1)$$

where t_R is the migration time of the solute, t_0 is the migration time of an unretained solute, and t_M is the migration time of the micelles. The capacity factor is a measure of the ratio of total moles of solute in the micellar phase versus those in the aqueous phase. As the velocity of the micellar phase slows and approaches zero, t_M becomes infinite, and the equation approaches that used in conventional chromatography.

To calculate the capacity factor, it is necessary to know the migration time not only of the analyte but also of the micelle and the EOF. Although there is no ideal marker in MECC, a very hydrophobic molecule, such as Sudan III, will spend most of its time partitioned in the micellar phase and can be used to mark the micellar phase; a very hydrophilic molecule, such as methanol, will spend very little time in the micellar phase and can be used to mark the EOF.

Resolution in MECC is given by[18]

$$R_s = \frac{N^{1/2}}{4}\left(\frac{\alpha - 1}{\alpha}\right)\left(\frac{k'_2}{1 + k'_2}\right)\left[\frac{1 - t_0/t_M}{1 + (t_0/t_M)k'}\right] \quad (5.2)$$

As with the capacity factor, as the velocity of the micellar phase approaches zero, the equation for resolution approaches that of classic chromatography.

5.3.2 Buffers

In MECC anionic surfactants are most frequently used, but cationic surfactants are also very popular. In addition, chiral surfactants, nonionic surfactants, zwitterionic surfactants, biological surfactants, or mixtures of each are finding increasing use. In all categories, variations in alkyl chain length will affect resolution or selectivity, as will changes in buffer concentration, pH, and temperature or the use of additives such as metal ions or organic modifiers. Typical surfactant systems used in MECC are shown in Table 5.3.

Surfactants suitable for MECC should (1) have enough solubility in the buffer to form micelles, (2) form a low-viscosity solution, and (3) form a solution that is homogeneous and UV transparent. The surfactant concentration should be kept below 200 mM because the viscosity of the buffer above that concentration, and therefore the current, becomes too high. The buffer concentration should be greater than 10 mM and higher than that of the surfactant in order to maintain a constant pH throughout the run.

The surfactant counterion is important in MECC because it affects the Kraft point, that is, the temperature above which the solubility of the surfactant increases sharply as a result of micelle formation; SDS has a lower Kraft point than potassium dodecyl sulfate and will therefore reach its CMC at a lower temperature. In MECC, many separation problems can be solved with standard MECC buffers and operating conditions; Table 5.4 provides a list of standard operating conditions.[19]

Figure 5.8 shows the separation of water-soluble vitamins by MECC, using a phosphate–borate buffer, pH 9.0, with 50 mM SDS.[20] The SDS concentration had the most dramatic effect on the migration times of vitamin B_1, probably due to ion-pair formation between the truly cationic vitamin and the polar group of the anionic surfactant. Nishi and Terabe[20] also investigated the effect of using N-lauroyl-N-methyltaurate in place of SDS and found a similar but less dramatic effect of surfactant concentration on migration time.

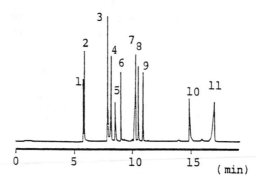

Figure 5.8 Separation of eleven water-soluble vitamins by MECC. Peaks: 1, pyridoxamine; 2, nicotinamide; 3 pyridoxal; 4, vitamin B_6; 5, vitamin B_2; 6, vitamin B_{12}; 7, vitamin B_2 phosphate; 8, pyridoxamine 5'-phosphate; 9, niacin; 10, vitamin B_1; 11, pyridoxal 5'-phosphate. Conditions: buffer, 50 mM SDS in 20 mM phosphate–borate buffer, pH 9.0; applied voltage, 20 kV; detection, UV absorbance at 210 nm. (Reprinted from Ref. 20 with permission.)

5.3.3 Practical Considerations

If the standard operating conditons do not provide the required separation, selectivity can be modified by changing a number of variables, including the nature of the surfactant and the aqueous phases. Altering the hydrophilic end of the surfactant has a dramatic effect since this is the end of the micelle that interacts with the solutes. Alternatively, a second surfactant can be added to form a mixed micelle. The addition of a nonionic surfactant to an ionic one decreases the migration time window so that the migration time of all the analytes decreases; nonionic chiral surfactants are often added to MECC buffers for the separation of enantiomers.

The addition of modifiers into the aqueous phase also affects selectivity. Cyclodextrins can be used in conjunction with the MECC buffer to provide selectivity for very hydrophobic analytes that would otherwise be almost totally incorporated into the micellar phase. In addition, they can be used for enantiomeric separations because of the chirality of the cyclodextrins themselves. The resolution between very hydrophobic compounds can also be improved by the addition of high concentrations of urea to the MECC buffer, which increases the solubility of hydrophobic compounds in water and breaks down hydrogen bonds in the aqueous phase. The addition of low concentrations (<20% v/v) of organic solvents, such as acetonitrile or 2-propanol, reduces the EOF and thereby expands the migration time window. Higher concentrations, however, can break down the micellar structure, so care should be taken. Finally, salts of certain metals, such as magnesium, zinc, or copper(II), can be added to enhance resolution of nucleotides. An optimization scheme for MECC separations is shown in Figure 5.9.

Figure 5.10 shows the separation of neutral flavinoids from plants.[21] Two surfactant systems were tested, one containing cetyltrimethylammonium bromide (CTAB) monitored at 350 nm and the other containing sodium cholate in combination with taurine monitored at 250 nm. The CTAB–MECC method resulted in hydrophobic interactions with the flavinoids as well as ion-pairing interaction with the flavinoids containing carboxyl groups. The cholate–MECC system resulted in hydrophobic interactions and ion repulsion of the flavinoids containing carboxyl groups. The authors concluded that the cholate system provided the best separations.

5.4 Capillary Gel Electrophoresis

Capillary gel electrophoresis (CGE) is a technique that has been developed to aid in the separation of macromolecules such as proteins and

Table 5.3 Typical Surfactant Systems Used in Micellar Electrokinetic Capillary Chromatography

Surfactant	Molecular formula	CMC $(10^{-3}\,M)^a$
Anionic		
Sodium decyl sulfate	$CH_3(CH_2)_9OSO_3^-Na^+$	8.1
Sodium dodecyl sulfate (SDS)	$CH_3(CH_2)_{11}OSO_3^-Na^+$	—
Sodium tetradecyl sulfate (STS)	$CH_3(CH_2)_{13}OSO_3^-Na^+$	2.1 (50°C)
Sodium dodecyl sulfonate	$CH_3(CH_2)_{11}SO_3^-Na^+$	—
Cationic		
Dodecyltrimethylammonium chloride (DTAC)	$CH_3(CH_2)_{11}N^+(CH_3)_3Cl^-$	16 (30°C)
Dodecyltrimethylammonium bromide (DTAB)	$CH_3(CH_2)_{11}N^+(CH_3)_3Br^-$	15
Cetyltrimethylammonium chloride (CTAC)	$CH_3(CH_2)_{15}N^+(CH_3)_3Cl^-$	—
Cetyltrimethylammonium bromide (CTAB)	$CH_3(CH_2)_{15}N^+(CH_3)_3Br^-$	0.92
Nonionic		
Octylglucoside		—
Triton X-100		0.24
Zwitterionic		
3-[3-(Cholamidopropyl)dimethylammonio]-1-propanesulfonate (CHAPS)		8
Chiral surfactants		
Digitonin		—
Sodium-N-dodecanoyl-L-valinate (SD Val)		5.7 (40°C)
Biological surfactants		
Bile salt surfactants		
Sodium cholate		13–15
Sodium taurocholate		10–15

^{a} Obtained using water as the solvent.

Table 5.4 Suggested Standard Operating Conditions for Micellar Electrokinetic Capillary Chromatography[a]

Buffer	50 mM SDS in 50 mM borate buffer (pH 8.5–9.0) or 50 mM SDS in 50 mM phosphate buffer (pH 7.0)
Capillary	50–75 μm I.D. × 20–50 cm (to detection window)
Applied voltage	10–20 kV (current below 100 μA)
Temperature	Ambient
Sample solvent	Water or methanol
Sample concentration	0.1–1 mg/ml
Polarity	Positive (injection side)
Detection	UV (200–210 nm)
Injection volume	<2 nl (1 mm)

[a] Adapted from Ref. 19 with permission.

nucleic acids, whose mass-to-charge ratios do not vary much with size. As illustrated in Figure 5.11, CGE provides a mechanism for separation on the basis of size differences, such that larger molecules are inhibited by the gel but smaller molecules pass through, largely unimpeded. Under well-controlled conditions, the mobility of the solute is inversely proportional to solute size. The first application of CGE to the analysis of macromolecules was reported in 1938 by Hjerten,[22] who separated proteins by capillary SDS-PAGE (sodium dodecyl sulfate–polyacrylamide gel electrophoresis). Since that time, many applications have been reported, including the separation of molecules such as oligosaccharides, polymerase chain reaction (PCR) products, and DNA restriction fragments.

5.4.1 Mechanism of Separation

In CGE, the capillary is filled with sieving material in the form of a gel or a viscous polymer (also referred to as a polymer network, entangled polymer solution, or physical gel). Two main theories have been proposed to describe the migration of a macromolecule through the sieving medium. The Ogston model[23] treats the solute as an undeformable spherical particle of radius R_g. Small solutes pass through the pores relatively uninhibited, but large solutes are restricted and their velocities reduced. In practice, however, large biopolymers do not always behave according to this model and are able to pass through much smaller pores than would be expected. The second model, known as the reptation model,[24–26] assumes biopolymers to be flexible molecules that wriggle through the pores head first. The smaller molecules still pass through the polymer network fastest, but the model allows for the fact that larger molecules pass through small pores

5.4 Capillary Gel Electrophoresis

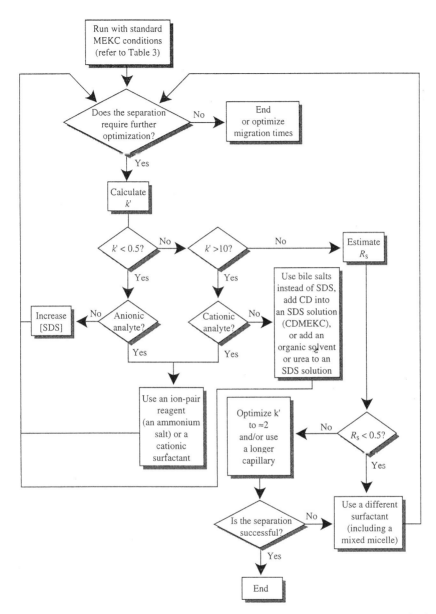

Figure 5.9 Optimization scheme for MECC separations. CD, cyclodextrin. For standard MECC conditions, see Table 5.4. (Reprinted from Ref. 19 with permission.)

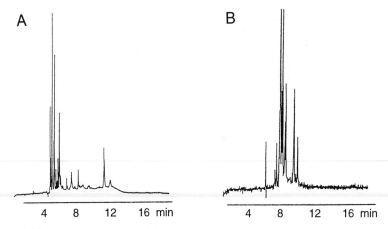

Figure 5.10 Separation of neutral flavinoids from leaves of rapeseed using (a) CTAB, and (b) using cholate in combination with taurine. (Reprinted from Ref. 21 with permission.)

and reach the detector. Readers are referred to Grossman[27] for in-depth information concerning the separation mechanisms in polymer networks and to Dubrow[28] for practical information on the technique.

5.4.2 Sieving Media

Many sieving materials have been introduced into capillaries for the purpose of separating biopolymers. The materials are classified as gels, which are cross-linked polymers, and polymer networks, which are not cross-linked. Cross-linked polyacrylamide, which is a widely used matrix, is usually polymerized *in situ* and not removed from the capillary. However, extreme care is required when making these capillaries to ensure that bubbles do not form and to prevent the gel from shrinking during polymerization. The major disadvantage of cross-linked polyacrylamide-filled gels is that polyacrylamide is hydrolytically unstable, and its use is therefore limited to a narrow pH range. In addition, owing to the rigid nature of the

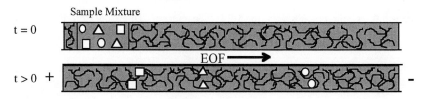

Figure 5.11 Schematic illustration of the separation mechanism in CGE.

5.4 Capillary Gel Electrophoresis

polyacrylamide matrix, dirty samples or clogging of the capillary ends may prevent reuse of the capillary.

Polymer networks (entangled polymer solutions) provide an alternative to cross-linked polyacryalmide gels. They are essentially low-viscosity gels and, as such, are more flexible than cross-linked gels. Polymer networks can be pumped readily into a capillary, and they are easy to regenerate and less costly than the cross-linked gels. Methylcellulose and derivatives are the most popular polymer networks, but both dextrans and polyethylene glycol have been used for protein separations. Cellulose derivatives are also attractive because the materials are commercially available and need only be dissolved and pumped into the capillaries. Figure 5.12 shows the separation of double-stranded DNA with monointercalating agents to allow laser-induced fluorescence detection.[29] The capillary was coated with linear polyacrylamide gel to eliminate the EOF and then filled with polyacrylamide according to the method by Hjerten.[30]

5.4.3 Practical Considerations

A major advantage of capillary gel electrophoresis is that resolution is maintained with increasing field strength, owing to efficient heat dissipation. Introduction of the sample into the gel-filled capillary is typically done electrokinetically (Chapter 6), because hydrostatic injection is not possible with capillaries blocked with gel. Typical injection times between 1 and 20 sec are used at field strengths between 100 and 400 V/cm.[28] Coated capillaries should be used to prevent the EOF from pumping the gel out of the capillary.

In semidilute solution, the average mesh size, ζ, of the polymer is

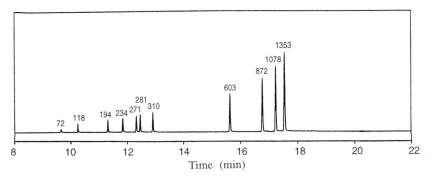

Figure 5.12 Electrophoretic separation of ΦX 174RF DNA *Hae*III digest using monointercalating dye (ethidium bromide). Conditions: buffer, 89 mM Tris, 89 mM boric acid, and 2 mM EDTA with 0.4% (w/v) (hydroxypropylmethyl)cellulose; electric field strength, 180 V/cm; injection, electromigration, 4 kV for 4 sec. (Reprinted from Ref. 29 with permission.)

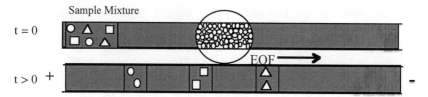

Figure 5.13 Schematic illustration of capillary electrochromatography.

dependent only on the polymer concentration;[24] the higher the concentration of the polymer solution, the smaller is the mesh size. At the crossover between dilute and semidilute polymer solution, however, the mesh size becomes dependent on the length of the polymer chain, instead.[24] To apply the polymers to the widest range of biopolymer separations, it is important to be able to vary the mesh size in solution. Although the mesh size can be decreased by increasing the concentration of polymer, the more concentrated the polymer, the more viscous it becomes. Therefore, the best way to manipulate the mesh size is to alter the length of the polymer chains.

Capillary gel electrophoresis has been employed both for molecular biology applications, such as PCR product analysis and antisense gene therapy, and for protein chemistry applications such as protein separations.

5.5 Capillary Electrochromatography

In capillary electrochromatography (CEC), the capillary is packed with a stationary phase that is capable of retaining solutes in a manner similar to column chromatography. Unlike the case in chromatography, however, there is no pressure-driven pump to force the solutes through to the detector; instead, the EOF is responsible for acting as an "electropump." The first report of an electroosmotic pumping system for liquid chromatography was published by Pretorius et al.[31] in 1974. Since that time, only a few articles have appeared, covering topics such as the control of linear velocity[32,33] and the effect of capillary diameter[34] and also providing techniques for producing packed columns.[33]

5.5.1 Mechanism of Separation

The most obvious advantage of an electroosmotic pumping system is the flow profile. Electrochromatography has the same pluglike profile that is characteristic of all other modes of CE. Although the pluglike flow is less perfect in packed columns than in open tubes, it is still more uniform than the laminar flow profile of a pressure-driven system. Consequently, the same capillary would be expected to provide higher efficiencies when used in CEC than when used in pressure-driven systems.[31,35,36]

The technique of CEC is illustrated in Figure 5.13. The separation

5.5 Capillary Electrochromatography

mechanism depends on both the electrophoretic mobility of the solutes and the nature of the packing material. Applications have included mostly standard reversed-phase type separations,[33] although ion-pair reversed-phase CEC has also been reported.[37]

5.5.2 Buffer Systems

The buffer system is a combination of buffer for electrophoresis and eluent for the particular chromatographic mode being employed. Figure 5.14 shows the separation of a group of neutral molecules using a capillary packed with a reversed-phase material.[38] The buffer was a mixture of 4 mM sodium tetraborate (pH 9.1) and acetonitrile (20:80, v/v). The separation was compared with a micro-HPLC separation in which the same capillary was used but the eluent was pressure driven. As can be seen in Figure 5.14, sharper peaks were obtained with the EOF-driven system.

5.5.3 Practical Considerations

There are several possible reasons for the slow development of capillary electrochromatography. The first is the difficulty involved in constructing the capillaries. However, if technology develops to the point where the capillaries can be packed easily and reproducibly, CEC columns will proba-

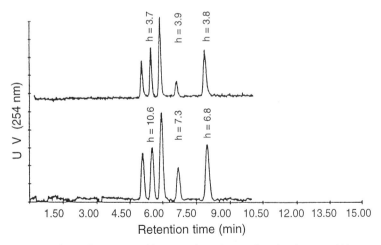

Figure 5.14 Electrochromatographic separation of neutral molecules on a 230 mm × 50 mm I.D. capillary packed with 3 mm Hypersil ODS. (A) Capillary electrochromatography, applied voltage 23 kV, (B) micro-HPLC, pressure 130 bar. Peaks, in order of elution, represent thiourea, benzylalcohol, benzaldehyde, benzene, and naphthalene. (Reprinted from Ref. 38 with permission.)

bly become more popular. A second reason is the separation mechanism. The popularity of CE modes such as CGE, CZE, and cIEF stems from the fact that larger molecules are not retained by the capillary, and therefore eluting them from the capillary is simple. In CEC, larger molecules will probably require gradient elution, as demonstrated in LC, and thus the advantage of simplicity that CE provides will be lost.

Nevertheless, CEC is still an attractive technique, because of low eluent consumption and small sample size requirements. It offers the advantage over LC of greatly increased plate counts and therefore the possibility of resolving closely related analytes not separable by LC. It is the least mature of the CE modes, but there is much interest in future development of CEC.

5.6 Chiral Capillary Electrophoresis

Chiral capillary electrophoresis is one of the most exciting new application areas for CE. Interest in chiral CE is driven by the promise of low-cost, free-solution methods for separating a broad range of chiral species. The various HPLC methods are comparatively complex and frequently require the use of expensive columns. Chiral CE provides much higher efficiencies, thereby allowing quantitation of trace levels of enantiomers, which is difficult in HPLC.

5.6.1 Mechanism of Separation

There are three possible approaches to the separation of chiral species by CE: (1) addition of chiral selectors to the buffer, (2) use of a chiral "stationary phase," and (3) precolumn derivatization. These correspond to the approaches in HPLC, and the separation mechanisms are described in Section 2.8. In the first approach, additives are added to CZE, CGE, or MECC buffers to effect the separation. In the second approach, chiral selectors can be immobilized on the capillary wall, although that is a difficult process. Alternatively, capillaries filled with enantiospecific packings can be employed for CEC. In the third approach, enantiomers are derivatized with chirally specific reagents prior to CZE or MECC. Addition of chiral selectors to the buffer is the most common approach.

5.6.2 Buffer Systems

The chiral selectors most commonly used as additives in the buffer can be divided into three main categories: inclusion systems [e.g., cyclodextrins (CDs) or crown ethers], enantioselective metal-ion complexes [e.g. copper(II)–L-histidine or copper(II)–aspartame], and optically active surfactants (e.g., chiral mixed micelles or bile acids). Cyclodextrins are the most widely reported, and they are used in low-pH buffers for the resolution of

5.6 Chiral Capillary Electrophoresis

basic compounds. The enantiomer that has the greatest interaction with the cyclodextrin will be retarded the most and will reach the detector last. Cyclodextrins can also be added to high-pH buffers, under which conditions the enantiomers will be negatively charged and will migrate away from the detector. In that case, then, the enantiomer that interacts most with the cyclodextrin will be the first to reach the detector, because its migration away from the detector will be retarded. Crown ethers can be employed to achieve chiral recognition of primary amines in a similar manner. Figure 5.15 shows the separation of four structurally complex basic drugs, each of which contains two aromatic rings.[39] A β-CD system was employed, and baseline resolution was obtained with most of the enantiomers.

Enantioselective metal chelation is a technique that has been applied to the separation of amino acid enantiomers. In the method, a transition metal–amino acid complex, such as copper(II)–aspartame, in which the full coordination of the complex has not been reached, is added to the buffer. The amino acid enantiomers are able to form ternary diastereomeric complexes with the metal–amino acid additive; if there are differences in stability between the two complexes, enantioselective recognition can be achieved.

The use of mixed micelles for chiral recognition was discussed in Section 5.3.3, using cyclodextrins. In addition to cyclodextrins, however, metal–amino acid complexes can also be used in a mixed mode arrangement. Bile salts are naturally occurring chiral surfactants that can be used as alternatives to, or in addition to, SDS for chiral recognition. In the presence of SDS, the migration times are faster. Table 5.5 shows initial operating conditions that can be used in chiral CE as a start to methods development.[40]

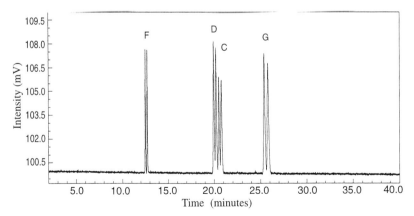

Figure 5.15 Separation of enantiomers in a β-cyclodextrin system. Conditions: buffer, 30 mM Tris–phosphate (pH 3) with 20 mM β-CD; separation voltage; 15 kV; samples, 0.1 mg/ml in buffer. (Reprinted from Ref. 39 with permission.)

Table 5.5 Suggested Operating Conditions for Chiral Capillary Electrophoresis[a]

Buffer	
Basic compounds	50 mM phosphate buffer (pH 2.5), 15 mM CD
Acidic compounds	50 mM borate (1 pH unit above the pK_a), 15 mM CD
Neutral compounds	50 mM borate, 50 mM SDS, 15 mM β-CD or 20 mM sodium taurocholate
Capillary	25–100 μm I.D. × 20–100 cm (to detection window)
Applied voltage	10–20 kV (current below 100 μA)
Temperature	Ambient
Polarity	Positive (injection side)
Detection	UV
Injection time	<20 sec

[a] Adapted from Ref. 40 with permission.

5.6.3 Practical Considerations

Selectivity can be manipulated by altering electrolyte properties, such as pH or ionic strength, and by incorporating into the electrolyte organic modifiers or other electrolyte additives. Table 5.6 lists the effects of various operating parameters.[40] If the chiral molecule contains no aromatic ring, α-CD or derivatized α-CDs can be used as a starting point. If the compound contains a single aromatic ring, β-CD or derivatized β-CDs may be more appropriate. Analytes with multiple rings are best separated with γ-CDs.

For acidic compounds, the pK_a should be established; the buffer should be adjusted so that it is at least one pH unit above the pK_a of the compound. Small neutral compounds are best separated using CDs, but bile acids may be better suited to the separation of the larger, more hydrophobic ones. Figure 5.16 shows methods development protocols for different classes of compounds.

5.7 Capillary Isoelectric Focusing

Capillary isoelectric focusing (cIEF) is used to separate zwitterionic compounds, such as peptides and proteins, on the basis of isoelectric point. The separation occurs as a result of a pH gradient that is established inside the capillary. It is important in cIEF to use coated capillaries to reduce or eliminate the EOF and to minimize solute adsorption; if the EOF is too swift, it could sweep the analytes past the detector before the separation is complete.

5.7.1 Mechanism of Separation

The pH gradient in cIEF is produced by the use of reagents, known as "carrier ampholytes," that are zwitterionic and are chosen so that the

5.7 Capillary Isoelectric Focusing

Table 5.6 Effects of Operating Parameters on Chiral Separations[a]

Variable	Range	Effect of increasing variable
Voltage	5–30 kV	Reduced analysis time, some loss in resolution
Current	5–250 mA	Reduced analysis time, some loss in resolution
Capillary length	20–100 cm	Increased analysis time, gain in resolution
Capillary bore	25–100 mm	Increased current, some loss in resolution
pH	1.5–11.5	Increased EOF, increased ionization of acids, reduced ionization of bases
Organic solvents	1–30% (v/v)	Gain or loss of resolution
Urea	1–7 M	Increased solubilization of hydrophobic solutes (and CDs)
Ion-pair reagent	1–20 mM	Can reduce or increase resolution
Amine modifiers	1–50 mM	Reduced surface charge, reduced peak tailing
CD type and size	—	Large impact on chiral selectivity
CD concentration	1–100 mM	Increased viscosity, reduced EOF, increased solute migration if complexation occurs
Viscosity	Various	Reduce EOF, longer migration times
Electrolyte concentration	5–200 mM	Increased resolution, increased current, lower EOF, solute ionization, reduced tailing
Cationic surfactant	1–20 mM	Reversal of EOF direction
Injection time	1–20 sec	Improved signal, some loss of resolution
Cellulose derivatives or polyvinyl alcohol	0.1–0.5%	Can improve resolution
Cationic surfactants	10–200 mM	Increased solubilization, longer migration times
Bile salt type	10–50 mM	Large impact on chiral selectivity

[a] Reprinted from Ref. 40 with permission.

span of their isoelectric points encompasses the isoelectric points of the analytes of interest. To perform a separation, the capillary is filled with a mixture of the sample and carrier ampholytes, which become uniformly distributed throughout the capillary, as illustrated in Figure 5.17. The compounds will be charged (cationic or anionic) depending on their pI values. When a voltage is applied, the ampholytes and solutes migrate to a position in the capillary where they become neutral (i.e., where pH = pI). Positively charged molecules migrate toward the cathode and negatively charged molecules migrate toward the anode. Complete separation of the ampholytes is never attained, as this would result in a discontinuity in the pH gradient. However, resolution between analytes with pI values as close as 0.02 pH units can be achieved.[41]

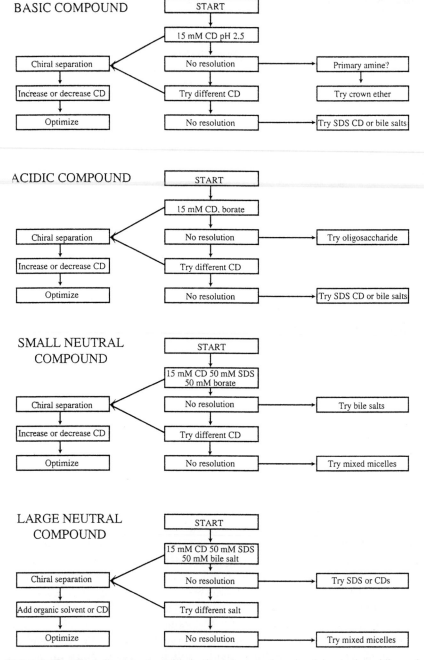

Figure 5.16 Flow diagrams for optimization of separations involving basic, acidic, and neutral compounds. (Reprinted from Ref. 38 with permission.)

5.7 Capillary Isoelectric Focusing

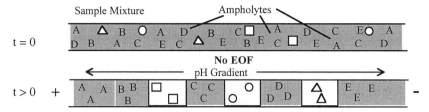

Figure 5.17 Schematic illustration of the separation mechanism in cIEF, where A–E represent different ampholytes.

5.7.2 Buffer Systems

The buffer system in cIEF is heterogeneous: the buffer in the reservoir surrounding the cathode (catholyte) is of a high pH, frequently 20 mM sodium hydroxide, whereas the buffer in the reservoir surrounding the anode (anolyte) is of a low pH, commonly 10–20 mM phosphoric acid. For narrow gradients (e.g., pH 6–8), weaker acids and bases such as glutamic acid and arginine, respectively, are used.[42]

5.7.3 Practical Considerations

In the most typical format, cIEF separations are performed in the absence of EOF in three distinct steps: the sample is introduced into the capillary, the analytes are "focused" or separated into distinct zones, and, finally, the zones are "mobilized" so that they pass by the detector. If the EOF is carefully controlled, it is also possible to perform cIEF separations in the presence of low EOF, in which case the mobilization step becomes unnecessary. To load the analytes, the desalted sample is usually mixed with a 1–2% (w/v) solution of the carrier ampholytes and the entire capillary is filled with the mixture. The procedure, however, requires the use of a relatively large amount of sample. By pretreating the capillary to eliminate the EOF (Section 4.3.3), it is possible to introduce smaller volumes of sample into the capillary.[43,44]

Following sample introduction the analytes are focused, or separated, into individual zones using the pH gradient. With a basic solution at the cathode and an acidic solution at the anode, when a voltage is applied to the system, the charged ampholytes and solutes migrate through the capillary until they reach a region where they become uncharged (i.e., reach their pI). In that region their mobility ceases. At the start of the process, the current is high as the zwitterions migrate rapidly toward their isoelectric points. As the molecules approach the pI values, the mobilities slow and the current drops, until a steady state is reached. Monitoring the current, therefore, is a convenient way to determine when the focusing is complete.

If a zwitterion diffuses out of the region of its isoelectric point, it becomes charged and rapidly migrates back into its own zone; thus, the separated zones remain sharp.

When the focusing is complete, the analytes must be removed from the capillary. If the separation occurs in the presence of a low EOF, the EOF will be sufficient to drive the solutes past the detector, and no extra steps are necessary. In the absence of any EOF, however, the analytes must be mobilized past the detector. There are two ways to mobilize the zwitterions: electrophoretic (pH) mobilization and hydrodynamic (pressure) mobilization. In hydrodynamic mobilization, a pump that was previously connected to the cathodic end of the capillary using a T-tube is started, and anolyte is pumped into the capillary at a low flow rate. The voltage used during the focusing step is maintained during the mobilization step to ensure that the zones remain sharp. In electrophoretic mobilization, a salt or a zwitterion is added to one of the two buffer reservoirs to effect pH changes in the capillary. Most commonly, the additive is placed in the cathodic reservoir (cathodic mobilization), and the voltage increased. If very acidic zwitterions are to be detected, however, anodic mobilization is employed. The same effect is achieved by replacing the acid in the anodic reservoir with base or by replacing the base in the cathodic reservoir with acid. In all cases, during mobilization the current increases gradually. Better resolution is always obtained during the early part of the electropherogram, owing to the effect on the pH gradient of mobilization. The resolution in cIEF, with respect to the difference in isoelectric points, is given by[41]

$$\Delta pI = 3 \left(\frac{D(dpH/dx)}{E(d\mu/dpH)} \right)^{1/2} \quad (5.3)$$

where D is the diffusion coefficient (cm^2/sec), E is the electric field (V/cm), dpH/dx is the change in buffer pH per unit length of capillary (as determined by the ampholyte range), and $d\mu/dpH$ describes the mobility–pH relationship. High resolution is obtained by the use of a high electric field and a narrow pH ampholyte range.

5.8 Capillary Isotachophoresis

In the electrolyte system used in capillary isotachophoresis (cITP), the sample zone migrates between a leading electrolyte at the front and a different, trailing electrolyte at the end. The leading electrolyte contains a coion with mobility greater than that of any of the analyte ions. The trailing electrolyte contains a coion with mobility that is lower than that of any of the analyte ions. In isotachophoresis, it is possible to analyze for anions or cations, but not both simultaneously. Analyses are usually performed in the constant-current mode.

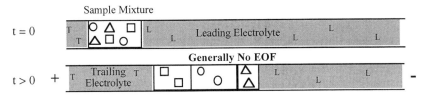

Figure 5.18 Schematic illustration of the separation mechanism in cITP.

5.8.1 Mechanism of Separation

Under the influence of an applied electric field, the components of the sample separate and form discrete zones, each of which is in contact with the zone of next highest and next lowest mobility. After complete separation, each zone contains only one individual component and the common counterion. The electrolyte coion is not present in the sample zones, and it is the sample ions, rather than the buffer coions, that are responsible for conduction of current through the sample zones, as illustrated in Figure 5.18. Once the system has reached equilibrium, all ions migrate with the same velocity, arranged in order of decreasing mobility. Because the analytes elute in order of decreasing mobility, a decreasing stepwise pattern of conductivity versus time is recorded by using a potential gradient detector.

The length of the analyte zone in cITP is a quantitative parameter and is related to the concentration of the analyte in that zone. The height of the step is a qualititative parameter that is characteristic of the analyte, and it is directly proportional to analyte mobility. Conductivity detection is the detection system generally used for isotachophoresis, although UV detection is employed occasionally in commercial instrumentation. Figure 5.19 shows a typical electropherogram obtained using cITP with conductivity detection.

5.8.2 Buffer Systems

Basic solutes with high pK_a values are separated using cITP in the cationic mode; acidic solutes are separated using the anionic mode. In the anionic mode, the buffer is selected so that the leading electrolyte contains an anion with an effective mobility that is higher than that of the solutes, and the terminating electrolyte has a lower mobility. In the cationic mode, the leading electrolyte contains a cation with a high effective mobility.

For anionic cITP, chloride is often selected as the leading ion; selection of the counterion is based on the operating pH, and amino acids or zwitterions are generally used. For cationic cITP potassium or ammonium are the leading ions of choice. Table 5.7 lists some buffer solutions for cITP at a series of pH values.

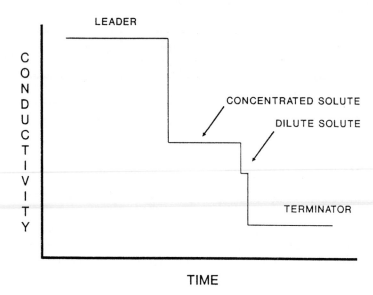

Figure 5.19 Two-component cITP separation with conductivity detection. (Reprinted from Ref. 6 with permission.)

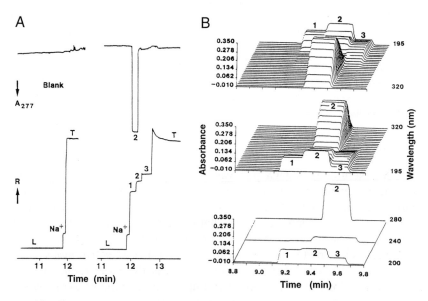

Figure 5.20 Separation of (1) ephedrine, (2) procaine, and (3) cycloserine by cationic cITP. (A) Ultraviolet absorbance and conductivity traces of blank and sample; (B) three-dimensional isotachopherogram. (Adapted from Ref. 45 with permission.)

5.8 Capillary Isotachophoresis

Table 5.7 Composition of Common Capillary Isotachophoresis Buffers[a]

Buffer composition	pH 3.3	pH 6.0	pH 8.8
Anionic cITP			
Leading ion	10 mM HCl	10 mM HCl	10 mM HCl
Leading counterion	β-Alanine	L-Histidine	Ammediol
Leading additive	0.2% HPME	0.2% HPMC	0.2% HPMC
Terminating ion	10 mM caproic acid	10 mM MES	10 mM β-alanine
Terminating counterion		Tris	Ba(OH)^2
Terminating pH		6.0	9.0
		pH 2.0	pH 4.5
Cationic cITP			
Leading ion		10 mM HCl	10 mM KOAc
Leading counterion			HOAc
Terminating ion		10 mM Tris	10 mM HOAc
Terminating counterion		HCl	
Terminating pH		8.5	

[a] Ammediol,2-amino-2-methyl-1,3-propanediol; HPMC, hydroxypropylmethylcellulose; MES, 2-(N-morpholino)ethane sulfonic acid; Tris, tris(hydroxymethyl)aminomethane. Reprinted from Ref. 6 with permission.

Figure 5.20 shows the separation of ephedrine, procaine, and cycloserine using cationic cITP. The leading electrolyte consisted of 10 mM potassium acetate and acetic acid (pH 4.75), and the terminating electrolyte consisted of 10 mM acetic acid; no additives were used.

5.8.3 Practical Considerations

Two important characteristics of cITP should be noted. First, compounds are separated by sharp boundaries. The sharply defined boundaries are an immediate result of the voltage gradient between the adjacent zones. The electric field is self-adjusting in order to maintain the constant velocity of the zones, and the zone with the lowest mobility experiences the highest field strength.

The second characteristic is that at equilibrium the concentration of the sample ions is related to the concentration of the leading ion. The concentration in each zone is constant and is determined by the concentration of the leading electrolyte. If the concentration of the leading electrolyte is high relative to the sample zones, the sample zones will narrow to approach the concentration of the leading ion. On the other hand, if the concentration of the leading electrolyte is low compared with the sample zones, the sample zones will broaden. Thus, concentration or dilution of the sample ions will occur, depending on the concentration of the analyte zone relative to that of the leading electrolyte. Capillary isotachophoresis, itself, is rarely used for the analysis of compounds of biological interest, but it is sometimes used as a preconcentrating step, prior to CZE.

5.9 Summary of Major Concepts

1. There are three distinct separation mechanisms operating in CE:

separation based on different velocities in an electric field,
separation based on different isoelectric points, and
separation based on different mobilities in a steady-state system.

The electrophoretic modes are all based on modifications of the three mechanisms.

2. There are seven capillary electrophoresis modes:

capillary zone electrophoresis (CZE),
capillary gel electrophoresis (CGE),
micellar electrokinetic capillary chromatography (MECC),
capillary electrochromatography (CEC),
chiral capillary electrophoresis (chiral CE),
capillary isoelectric focusing (cIEF), and
capillary isotachophoresis (cITP).

3. Capillary zone electrophoresis is the simplest and most widely used mode in CE. Separations in CZE are based on mobility differences between the analytes in an electric field. These differences are dependent on the size and charge-to-mass ratio of the analyte ion.

4. The separation mechanism in MECC is similar to that in CZE, except that micelles are added to the buffer system. The MECC technique is applied, usually but not exclusively, to the separation of neutral species.

5. The CGE technique has been developed to aid in the separation of macromolecules such as proteins and nucleic acids, whose charge-to-mass ratios do not vary much with size. Separation is based on size differences as molecules pass through a sieving medium.

6. In CEC, the capillary is packed with a stationary phase; thus, the separation mechanism is a combination of that found in LC and CZE.

7. The separation mechanisms in chiral CE are similar to those encountered in chiral LC. One approach is to use a chiral stationary phase, another is to add chiral selectors to the buffer, and the third is to conduct precolumn derivatization.

8. In cIEF, zwitterionic compounds such as peptides and proteins are separated on the basis of their isoelectric points.

9. In cITP, the sample zone migrates between a leading electrolyte and a trailing electrolyte. It is possible to analyze for anions or cations, but not for both simultaneously.

References

1. Strickland, M., and Strickland, N., *Am. Lab.* **November**, 60 (1990).
2. Honda, S., Suzuki, S., Nose, A., Yamamoto, K., and Kakehi, K., *Carbohydr. Res.* **215**, 193 (1991).
3. Kaneta, T., Tanaka, S., and Yoshida, H., *J. Chromatogr.* **538**, 385 (1991).
4. Heiger, D. N., "High Performance Capillary Electrophoresis—An Introduction." p. 46. Hewlett-Packard Company, France, 1992.
5. McLaughlin, G. M., Nolan, J. A., Lindahl, J. L., Palmieri, R. H., Anderson, K. W., Morris, S. C., Morrison, J. A., and Bronzert, T. J., *J. Liq. Chromatogr.* **15**(6/7), 961 (1992).
6. Weinberger, R., "Practical Capillary Electrophoresis." Academic Press, San Diego, 1993.
7. Mosher, R. A., *Electrophoresis* (*Weinheim*) **11**, 765 (1990).
8. Weston, A., Brown, P. R., Jandik, P., Jones, W. R., and Heckenberg, A. L., *J. Chromatogr.* **593**, 289 (1992).
9. Weston, A., Brown, P. R., Heckenberg, A. L., Jandik, P., and Jones, W. R., *J. Chromatogr.* **602**, 249 (1992).
10. Jandik, P., Jones, W. R., Weston, A., and Brown, P. R., *LC-GC* **9**, 634 (1991).
11. Guzman, N. A., and Hernandez, L., in "Techniques in Protein Chemistry" (T. E. Hugli, ed.,) Academic Press, Orlando, Florida, 1989.
12. Nielsen, R. G., Riggin, R. M., and Rickard, E. C., *J. Chromatogr.* **480**, 393 (1989).
13. Nashabeh, W., and El Rassi, Z., *J. Chromatogr.* **536**, 31 (1991).
14. Nguyen, A. L., Luong, J. H. T., and Masson, A., *Anal Chem.* **62**, 2490 (1990).
15. Stromqvist, M., *J. Chromatogr.* **667**, 304 (1994).
16. Cohen, A. S., Terabe, S., Smith, J. A., and Karger, B. L., *Anal. Chem.* **59**, 27 (1987).
17. Terabe, S., Otsuka, K., Ichikawa, S., Tsuchiya, A., and Ando, T., *Anal. Chem.* **56**, 111 (1984).
18. Terabe, S., Otsuka, K., and Ando, T., *Anal. Chem.* **57**, 834 (1985).
19. Terabe, S., "Micellar Electrokinetic Chromatography." Beckman Instruments, Fullerton, California, 1993.
20. Nishi, H., and Terabe, S., *Electrophoresis* (*Weinheim*) **11**, 691 (1990).
21. Bjergegaarf, C., Michaelsen, S., Mortensen, K., and Sorensen, H., *J. Chromatogr.* **652**, 477 (1993).
22. Hjerten, S., *J. Chromatogr.* **270**, 1 (1983).
23. Ogston, A. G., *Trans. Faraday Soc.* **54**, 1754 (1958).
24. de Gennes, P. G., "Scaling Concepts in Polymer Physics." Cornell Univ. Press, Ithaca, New York, 1979.
25. Lerman, L. S., and Frisch, H. L., *Biopolymers* **26**, 995 (1982).
26. Doi, M., and Edwards, S. F., *J. Chem. Soc., Faraday Trans. II* **79**, 1789 (1978).
27. Grossman, P. D., in "Capillary Electrophoresis: Theory and Practice," (Paul D. Grossman and Joel C. Colburn, eds.) Chapter 8. Academic Press, San Diego, 1996.
28. Dubrow, R. S., in "Capillary Electrophoresis: Theory and Practice," (Paul D. Grossman and Joel C. Colburn, eds.) Chapter 5. Academic Press, San Diego, 1992.
29. Kim, Y., and Morris, M. D., *Anal. Chem.* **66**, 1168 (1994).

30. Hjerten, S. J., *J. Chromatogr.* **347**, 191 (1985).
31. Pretorius, V., Hopkins, B. J., and Schieke, J. D., *J. Chromatogr.* **99**, 23 (1974).
32. Stevens, T. S., and Cortes, H. J., *Anal. Chem.* **55**, 1365 (1983).
33. Knox, J. H., and Grant, I. H., *Chromatographia* **32**, 317 (1991).
34. Bruin, G. J. M., Tock, P. P. H., Kraak, J. C., and Poppe, H., *J. Chromatogr.* **517**, 557 (1990).
35. Knox, J. H., *Chromatographia* **26**, 329 (1988).
36. Jorgenson, J. W., and Lukacs, K. D., *HRC & CC, J. High Resolut. Chromatogr. Chromatogr. Commun.* **8**, 407 (1985).
37. Pfeiffer, W., and Yeung, E. S., *J. Chromatogr.* **557**, 125 (1991).
38. Yan, C., Schaufelberger, D., and Erni, F., *J. Chromatogr.* **670**, 15 (1994).
39. Nielen, M. W. F., *Anal Chem.* **65**, 885 (1993).
40. Rogan M. M. and Altria K. D., "Introduction to the Theory and Applications of Chiral Capillary Electrophoresis." Beckman Instruments, Fullerton, California, 1993.
41. Righetti, P. G., "Laboratory Techniques in Biochemistry and Molecular Biology." (T. S. Work and R. H. Burdon, eds.). Elsevier, Amsterdam, 1983.
42. Mazzeo, J. R., and Krull, I. S., *Anal. Chem.* **63**, 2852 (1991).
43. Thormann, W., Caslavska, J., Molteni, S., and Chmelik, J., *J. Chromatogr.* **589**, 321 (1992).
44. Chen, S.-M., and Wiktorowicz, J. E., *Anal. Biochem.* **206**, 84 (1992).
45. Gebauer, P., and Thorman, W., *J. Chromatogr.* **545**, 299 (1991).

CHAPTER 6

Instrumentation for Capillary Electrophoresis

6.1 Introduction

A capillary electrophoresis system is comparatively simple. The basic components (Fig. 6.1) include the power supply which provides the high voltage necessary for the separation, the capillary in which the separation takes place, the detector which determines the sensitivity of the separation, and the data acquisition system which records the electropherogram. Some instruments also perform fraction collection. The final electropherogram looks similar to a chromatogram obtained from HPLC.

To make an injection, the source electrolyte vial is removed and a sample vial is put in its place. A small quantity of sample is introduced into the end of the capillary by either electromigration or hydrostatic injection, and then the source electrolyte vial is returned to the original position. The power supply is switched on, and the analyte ions migrate toward the detector. As the ions pass by the detector they are visualized, and the signal is recorded on the data acquisition device. The majority of the literature using CE describes the application of UV or fluorescence detection since these were the only detectors offered by the early commercial instrument manufacturers. "Next generation" instruments offer conductivity and mass spectrometry detection in addition to fluorescence and UV, so applications using those detectors are no longer limited to university research departments.

One of the major differences among commercially available instruments is the approach to cooling the capillary. The options include liquid cooling, forced air–nitrogen convection, or no mechanical circulation. Liquid cooling provides the most control, but forced air–nitrogen is sufficient for the purposes of reproducibility. The main disadvantage of liquid cooling is the inconvenience associated with changing the capillaries, which are housed within a specially designed cartridge. Capillary cooling is of most importance when using high ionic strength buffers and high voltages; the

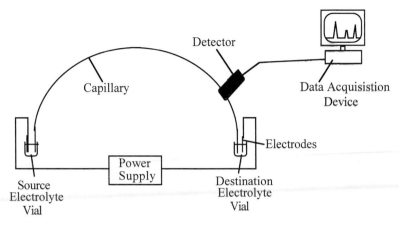

Figure 6.1 Schematic diagram of a capillary electrophoresis instrument, not drawn to scale.

analysis of small inorganic species, for example, requires the use of low ionic strength buffers, and capillary cooling is of less significance than for peptide analyses.

6.2 High-Voltage Power Supply

In CE, a high-voltage DC power supply provides the electrical field necessary to establish the electroosmotic flow of the bulk solution as well as electromigration of the charged analytes. Most power supplies provide from −30 kV to +30 kV with current levels of about 200 to 300 μA. As it is necessary to make the detector end either anodic or cathodic depending on the application, it must be possible to switch polarity.

The most common operating mode in CE is constant voltage, and stable voltage regulation is imperative to maintain good migration time reproducibility. Occasionally constant current or constant power may provide better reproducibility, as in situations where the ionic strength of one sample in a batch of samples differs significantly. Constant current is more common in cITP than is constant voltage. Thus, the power supply should be able to provide all three modes of operation. The use of pressure in addition to one of the three modes mentioned above provides some advantage during the separation. The analysis time is reduced if pressure is also applied. In addition peak efficiency is increased up to a pressure of about 0.2 psi.

Field programming is the ability to run voltage, current, or power gradients, and it is another feature that is often useful. Although field gradients are not common, they have been used to slow down the migration of an analyte of interest prior to fraction collection. In addition, field programming has been used to decrease the analysis time of a sample mixture when the time between the last group of peaks is large.

6.3 Sample Injection

In CE, only small quantities of sample can be introduced onto the capillary if the high efficiencies characteristic of the technique are to be maintained, as discussed in connection with electromigration dispersion in Section 4.3.4. In general, the sample length should be less than 2% of the total capillary length. Although this can be an advantage for applications with limited volumes of sample, it can be a problem from the point of view of detection.

6.3.1 Hydrostatic Injection

Sample is introduced into the capillary by hydrostatic injection (gravity, pressure, or vacuum) or electromigration injection. Hydrostatic injection, which is the most common approach, can be achieved by (1) raising the sample vial a given distance above the level of the destination vial for a predetermined time (gravity), (2) applying pressure to the sample vial (pressure), or (3) applying a vacuum to the destination end of the capillary (vacuum). The three sample introduction mechanisms are illustrated in Figure 6.2.

With hydrostatic injection mechanisms, injection reproducibility can be better than 1–2% RSD. The volume of sample loaded is a function of the capillary dimensions, the viscosity of the buffer, the applied pressure, and the time, and it can be calculated using

$$\text{Volume} = \frac{\Delta P d^4 \pi t}{128 \eta L} \tag{6.1}$$

where ΔP is the pressure difference across the capillary, d is the capillary inner diameter, t is the time, η is the buffer viscosity, and L is the total capillary length. For gravity injections, ΔP is given by

$$\Delta P = \rho g \Delta h \tag{6.2}$$

where ρ is the buffer density, g is the gravitational constant, and Δh is the height differential of the reservoirs.

6.3.2 Electromigration Injection

Electromigration injection is performed by replacing the source electrolyte vial with the sample vial and applying voltage, as illustrated in Figure 6.3. The advantage of this approach over the others is that more sample

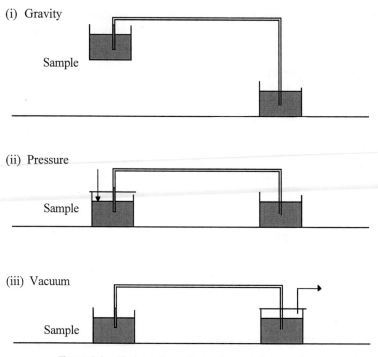

Figure 6.2   Hydrostatic methods of sample introduction.

can be introduced, thereby improving the detection limits. The disadvantage is that the sample slug that enters the capillary is not representative of the sample, because the higher mobility ions will enter at a greater rate than the lower mobility ions. The quantity of sample introduced by electromigration injection can be calculated:

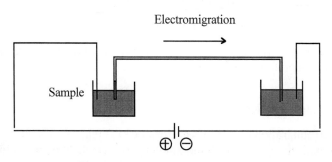

Figure 6.3   Electromigration injection.

6.3 Sample Injection

$$Q = \frac{(\mu_e + \mu_{eo})V\pi r^2 Ct}{L}. \tag{6.3}$$

From Eq. (6.3), sample loading is dependent on the EOF, the sample concentration, and the sample mobility. μ_e = electrophoretic mobility of the analyte, μ_{eo} = EOF mobility, V = voltage, r = capillary radius, C = analyte concentration, t = time, L = capillary total length.

6.3.3 Sample Concentration

If greater sample loading is required, the sample can be "stacked" during the injection process. Sample stacking is a concentrating effect of the analyte zone and occurs when the sample is dissolved in a more dilute solution than the buffer. Figure 6.4 shows the effect on peak height of using hydrostatic injection, electromigration injection where the sample is dissolved in the buffer, and electromigration injection where the sample is dissolved in water, for a mixture of peptides.[1]

Sensitivity can also be enhanced by the use of field-amplified sample injection (FASI).[2] In the FASI technique, sample is injected into the capil-

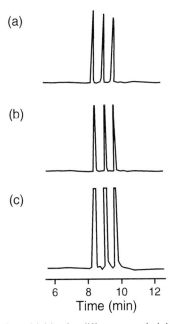

Figure 6.4 Comparison of sensitivities for different sample injection techniques using a 50 mg/ml peptide mixture: (a) hydrostatic injection, (b) electromigration injection where the sample is dissolved in buffer, and (c) electromigration injection where the sample is dissolved in water. (Adapted from Ref. 1 with permission.)

lary hydrostatically until the capillary is almost filled, as illustrated in Figure 6.5a. The sample is then focused by the application of a high voltage in the reverse direction, which removes water from the capillary (Fig. 6.5b). When the current drops to about 95% of the original value, the voltage is switched back to the normal configuration, and separation is performed (Fig. 6.5c).

6.4 Capillaries

Fused silica capillaries are the most popular capillaries and are commercially available with internal diameters ranging from 10 to 200 μm, although the most commonly used capillary dimensions are 25, 50, and 75 μm I.D. and 350–400 μm O.D. To minimize distorting refractive index effects and light scattering with on-column optical detection, the ratio of the outer

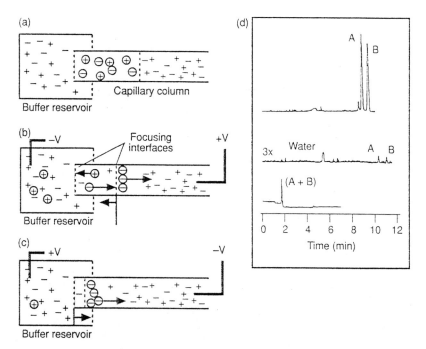

Figure 6.5 Chemical enhancement by field amplification: (a) sample is introduced into the capillary by hydrostatic injection, (b) sample is focused by the application of a high voltage in the reverse direction, and (c) voltage is switched to the normal configuration for separation. (d) electropherogram showing improvement in sample stacking of a large injection volume. Top: 35 cm sample plug with sample buffer removed; middle: conventional stacking with 1 cm sample plug; bottom: same as top without removal of buffer. (A) phenylhydantoin (PTH)-aspartic acid, (B) PTH-glutamic acid. (Adapted from Ref. 2 with permission.)

6.4 Capillaries

diameter to the inner diameter is relevant and should be equal to 4.7.[3] Thus, a capillary with an inner diameter of 75 μm and an outer diameter of 353 μm will give minimal ellipticity of the ouput beam, as will a capillary with 50 μm I.D. and 235 μm O.D.

Capillaries may be divided into several categories, depending on the application for which they are to be used. Broadly speaking, untreated or uncoated capillaries are used for the analysis of inorganic or low molecular weight organic species such as anions,[4-8] cations,[9-13] and organic acids, as well as for smaller molecules of biological interest, such as amino acids,[14] peptides,[15] and some proteins.[16,17] For the analysis of most larger proteins, and for isoelectric focusing applications, it is important to eliminate the EOF as well as interactions between the protein and the capillary wall. In general, it can be assumed that separations by MECC will be performed in uncoated capillaries, cIEF will be performed in coated capillaries, CZE can be achieved in either (depending on the analysis), and CGE will be performed in gel-filled capillaries. The cITP technique is generally performed in Teflon capillaries, although separations in coated and uncoated fused silica capillaries have been reported.

6.4.1 Untreated Fused Silica Capillaries

Untreated fused silica capillaries are most commonly used for the CE or MECC analysis of inorganic and smaller organic molecules, but other applications involve carbohydrates, amino acids, myotoxins, polyamines, proteins, peptides, and antidepressants. The advantage of using untreated capillaries is the ease of use, as no complicated capillary pretreatment is necessary. In addition, because high-quality fused silica tubing is available, the internal surface is fairly predictable, and so the separations are fairly predictable. The disadvantages, however, include the possibility of interaction between the solute and the capillary wall. Because the silica capillary wall possesses hydroxyl groups on the interior surface, larger, less rigid solutes such as proteins and large peptides tend to adsorb to the capillary wall.

In uncoated capillaries, the problem of solute interaction is especially a problem with basic proteins, which carry a significant positive charge when operating in the typical pH range used for CE. The charge on the protein results in strong coulombic interactions with the acidic silanols, which can, on occasion, lead to irreversible protein adsorption. The problem can be addressed in two ways: the first involves varying the pH of the buffer relative to the isoelectric point of the protein to change the net charge on the protein, whereas the second involves changing the characteristics of the wall by adding a component to the buffer that will preferentially adsorb to the wall or in some other way modify the wall.

6.4.2 Functionalized Fused Silica Capillaries

Coating the interior surface of the capillary with an uncharged hydrophilic material reduces or eliminates electroosmotic flow. To be useful for routine analysis, however, the coating must be stable under normal operating conditions. If the capillary coating deteriorates over time, the efficiency and reproducibility of the separation will be compromised.

Deactivation of the surface of the fused silica is the most common approach to reduce adsorption of basic proteins to the capillary wall, and a number of modifying techniques have been described. Although some of these modification procedures have been developed to inhibit the EOF,[18,19] others have been designed to keep a certain level of EOF to permit the analysis of oppositely charged species in a single run.[20–22] Capillary coatings have included methyl cellulose,[18,23] hydroxypropyl cellulose,[24] non-crosslinked polyacrylamide,[25,26] epoxydiol,[27] maltose,[27] polyethylene glycol,[21,28,29] polyethyleneimine,[22] aryl pentafluorol,[20] and LC reversed-phase (e.g., C_{18}, C_8, C_2) coatings.[30,31] The characteristics of some polymers used as coatings are listed in Table 6.1;[32] polyacrylamide and polyethylene glycol are two of the most commonly cited polymers. Although adsorption is decreased by the application of these coatings, drawbacks associated with surface deactivation include inconsistent performance from column to column and short lifetimes of the coatings. For a good description of the various bonded phases, interested readers are referred to Li.[32]

6.4.3 Extended Path-Length Capillaries

In CE detection is performed on-column, so the detection cell is defined by the diameter of the capillary. Because sensitivity is proportional to the path length, sensitivity of absorbance methods is limited. Thus, capillary

Table 6.1 Characteristics of Polymers Used as Coatings[a]

Polymer	Hydrophobicity/ hydrophilicity	Hydrogen-bonding capacity
Polyethylene	Very hydrophobic	None
Poly(ethylene glycol) (Superox)	Weakly hydrophilic	Low
Poly(ethylene–propylene glycol) (Ucon)	Weakly hydrophilic	Low
Polyacrylamide	Hydrophilic	High
Poly(hydroxypropyl cellulose) (HPC)	Strongly hydrophilic	High

[a] Reprinted from Ref. 32 with permission.

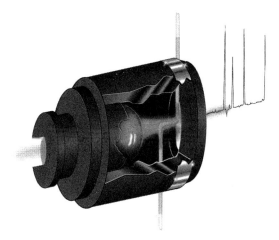

Figure 6.6 Illustration of the Z-cell and the light path through it. (Reprinted from Ref. 34 with permission.)

geometries have been developed to address this problem. In the Z-cell[33] shown in Figure 6.6,[34] the capillary is bent into a Z shape, resulting in a longitudinal optical path length of 3–4 cm and sensitivity enhancements of up to 6 times. In the bubble cell[35] shown in Figure 6.7,[36] a bubble is blown into a capillary in the region of the detector, producing a cell with an extended path length. Both the Z-cell and the bubble cell capillaries are commercially available.

6.5 Detection

Many of the detectors used in HPLC have been adapted for use in CE. This has been a major challenge, as the highly efficient separations afforded by CZE are a direct result of employing extremely narrow separation columns. Narrow inner diameter capillaries are required to provide effective dissipation of the heat generated by the passage of electrical

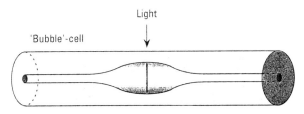

Figure 6.7 Illustration of a capillary in which a bubble cell has been blown. (Reprinted from Ref. 36 with permission.)

current through the capillary. However, the small capillary dimensions and the consequently small sample loadings create a system in which sensitive detection of solute zones, without introducing zone dispersion, is a real issue. This chapter discusses the more commonly used detection schemes and the separations for which they are applicable. Details on detector design are not provided here; instead, readers are referred to several comprehensive books, and chapters of books, dedicated to CE detectors.[37-39] The theory of the detection method was presented in Chapter 3 on HPLC detection. This chapter presents the application of these detectors to CE.

6.5.1 Detector Properties

An ideal on-line detector for HPLC or CE has versatility, high sensitivity, continuous monitoring of the column effluents, low noise level, wide linearity of response, stable baseline, insensitivity to flow rate and temperature changes, and response to all types of compounds. It should be rugged and not too expensive. The detector should be able to measure accurately a small peak volume without increasing its volume appreciably. Of these parameters, the terms noise, sensitivity, and linearity are typically used in describing detector performance.

6.5.2 Absorbance Detectors

Absorbance detectors are nondestructive and respond only to those substances that absorb light at the wavelength of the source light. They are usually classified as selective or solute property detectors Detectors that measure only in the range 190–350 nm are ultraviolet (UV) absorption detectors, whereas those that measure in the region 350–700 nm are visible (Vis) detectors. Detectors that span the range 190–700 nm are known as UV/Vis detectors. The UV and UV/Vis detectors are the most widely used detector types for both HPLC and CE, owing to both the relative insensitivity of the detector to temperature and gradient changes and to the great number of compounds that absorb light in the UV range.

(i) Direct Absorbance Detection

There are three aspects to consider in the design of an on-column UV detector cell for CE. Detection is performed on-column, most commonly in fused silica capillaries, because they have a UV cutoff near 170 nm while that of glass is approximately 300 nm. The protective coating on the silica capillary is removed in the region of the detector, to expose the on-column cell. Light should pass only through the inner diameter of the capillary. If a large amount of light passes through the rim of the capillary, the signal will be very sensitive to refractive index changes of the solution, and the signal-to-noise ratio (S/N) and the linear range of detection will be greatly

6.5 Detection

reduced.[40] The aperture should have a width, along the axis of the capillary, that keeps the efficiency loss within a predefined limit (<1 sec of the peak). Finally, installation and removal of capillaries should be convenient and accurate.

In 1986, Foret et al.[41] described an on-line UV absorbance detector that employed a commercial photometer and optical fibers in direct contact with the outer walls of the separation capillary. The optical fibers (200 μm I.D. fused silica core) conducted the light beam perpendicularly across the migrating zones; one fiber was connected to a mercury lamp to serve as the illumination source, and the other directed light to a photomultiplier tube for detection. The detector was found to be linear in the range of 10^{-5} to 10^{-3} M ($r = 0.994$ for 10 measurements), with detection limits of 1×10^{-5} M for picric acid ($S/N = 2$).

In 1988, Brownlee and Compton[42] used a photodiode array detector to analyze minute quantities of complex biological compounds. Unfortunately, the small diode array employed limited detection to fewer than 20 wavelengths. Kobayashi et al.[43] used a full 512-element diode array to provide coverage of the 200–380 nm UV range for the analysis of complex mixtures of aromatic compounds and vitamins. That system takes advantage of full UV spectrum scanning with the concomitant benefits of contour plotting of spectra for rapid and accurate characterization of complex samples.

Figure 6.8 shows the use of a photodiode array detector to verify the identity of some antibiotics.[44] The sample contained six antibiotics, all of which absorb in the region 198–200 nm. Only two antibiotics, namely, nystatin and tylosin tartrate, were found to absorb in the region 306–308 nm. Thus, the photodiode array detector was able to verify the peak assignment without the need for further injections.

(ii) Indirect Absorbance Detection

The first report demonstrating the feasibility of indirect detection in CE was published in 1987 by Hjerten et al.[45] who employed indirect UV absorbance detection for the analysis of both inorganic ions and organic acids. The UV-background-providing electrolyte was 25 mM sodium veronal, pH 8.6, and detection was monitored on-column at 225 nm. In 1990, the first separation of alkali, alkaline earth, and lanthanide metals was reported by Foret et al.[46] Indirect UV detection at 220 nm was employed to detect 14 metals in 5 min, with baseline resolution achieved between all but two of the components. The baseline showed a reproducible upward drift between 1 and 3 min. The UV-absorbing component of the electrolyte was creatinine, with α-hydroxyisobutyric acid introduced to complex with the lanthanides and improve resolution.

Indirect UV is the most common mode for the detection of inorganic anions and cations by CE. Pyromellitic acid or chromate are the most

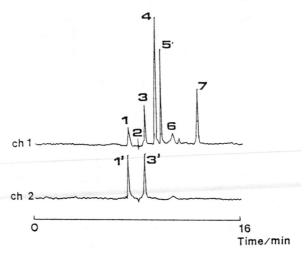

Figure 6.8 Electropherogram of six antibiotics. Conditions: buffer, 50 mM phosphate/0.1 M borate (pH 7.06); capillary, 50 cm × 50 μm I.D. fused silica; voltage, 15 kV; detection, UV absorbance at 198–200 nm (channel 1) and 306–308 nm (channel 2). Peaks: 1, tylosin tartrate; 2, methanol; 3, nystatin; 4, amoxicillin; 5, ampicillin; 6, chlortetracycline; 7, penicillin G. (Reprinted from Ref. 44 with permission.)

common electrolytes for high-mobility anions, with an additive such as tetradecyltrimethylammonium bromide (TTAB) present to reverse the EOF. Benzoate is more appropriate for lower mobility anions if peak symmetry is to be maintained. The electrolytes for cations are usually amines, since those compounds are the most commonly available UV-absorbing cationic species. Other compounds are available, however, and copper(II) sulfate has been used with some success as has dimethyldiphenylphosphonium hydroxide (DDP). Complexing agents, typically organic acids, must be added in order to resolve inorganic cations, particularly transition metals and lanthanides, and 18-crown-6 ether is used to separate ammonium from potassium. Figure 6.9 shows the indirect UV detection of inorganic cations by CE using an amine as the UV-absorbing background electrolyte.[10] The complexing agent used to separate the analytes was tropolone.

6.5.3 Fluorescence Detectors

Fluorescence is a specific type of luminescence. Fluorescent molecules absorb light at one wavelength, and then reemit it, essentially instantaneously, at a longer wavelength (lower energy). Fluorescence is expected in molecules that are aromatic or contain multiple conjugated double bonds with a high degree of resonance stability. In either case, the compound will possess delocalized π electrons that can be placed in low-lying excited singlet states. Polycyclic aromatic systems, which possess a great number of delocalized π electrons, are usually more fluorescent than benzene and benzene derivatives. Electron-donating groups present on the ring, such as $-NH_2$, $-OH$, $-F$, $-OCH_3$, and $-N(CH_3)_2$, tend to enhance fluorescence, whereas electron-withdrawing groups, such as $-Cl$, $-Br$, $-I$, $-NO_2$, and $-CO_2H$, tend to decrease or quench fluorescence. Thus, aniline fluoresces, but nitrobenzene does not. Molecular rigidity is also important for fluorescence, and in general, given a series of aromatic compounds, those that are the most planar, rigid, and sterically uncrowded are the most fluorescent. Occasionally, molecules with nonbonding electrons, such as amine groups, will fluoresce, but usually a delocalized p system must be part of the molecule.

As with absorbance detection, fluorescence detection is easily adapted for use in CE. An advantage of fluorescence over absorbance detection is that sensitivity is comparatively independent of path length. In fluorescence detection, excitation radiation from an arc lamp or a laser is focused onto a section of the capillary where the polyimide coating has been removed. The resulting fluorescence is collected at an angle of 90° relative to the excitation beam through a system of lenses or an optical fiber. Fluorescence detection has been applied to the analysis of solutes in capillaries with inside diameters ranging down from 75 to 25 μm.[47]

Unlike the case with UV detection, few commercial HPLC fluorescence detectors may be suitably modified to allow efficient use with small capillaries.[48] Consequently, custom-built systems are the rule when fluorescence detection is employed with capillary electrophoresis. Because of superior focusing capabilities, which allow the excitation energy to be more effec-

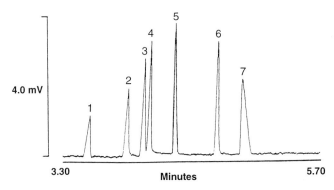

Figure 6.9 Indirect UV detection of inorganic cations by CE. Conditions: electrolyte, UVCat2 with tropolone as the complexing agent (pH 4.4); detection, indirect UV absorbance at 190 nm; capillary, 60 cm $\times$ 75 μm I.D. fused silica. Peaks: 1, Potassium; 2, Barium; 3, Strontium; 4, Calcium; 5, Sodium; 6, Magnesium; 7, Lithium. (Reprinted from Ref. 10 with permission.)

tively applied to very small sample volumes, and better monochromicity, which reduces stray light levels, lasers have been employed as the excitation source for fluorescence detectors used with capillary electrophoresis. Figure 6.10 shows the laser-induced fluorescence detection of hemoglobin (Hem A), carbonic anhydrase (CAH), and methemoglobin (Met) in a hemolysate of human erythrocytes.[49] The mass limits of detection ($S/N = 2$) for CAH and HemA were found to be 0.2 and 8 amol, respectively (6 fg and 0.5 pg).

Although on-column fluorescence detection can provide excellent detection limits, the technique is less versatile than UV detection because many solutes of interest do not exhibit native fluorescence and must be derivatized with some type of fluorophore. Consequently, the literature contains many examples of capillary electrophoresis separations of fluorescence-labeled solutes such as dansylated amino acids.[50,51]

An alternative to derivatization of nonfluorescent compounds is to perform indirect fluorescence detection. The procedure is performed on-column, by incorporating a fluorescent ion into the electrolyte. When ionic analytes interact with the fluorophore, the result is either displacement of the fluorophore or ion pairing with it.[47] Kuhr and Yeung[51] explored indirect laser-induced fluorescence detection using 1.0 mM salicylate as the fluorophore in the electrolyte; they analyzed 10 amino acids and obtained detection limits on the order of 10^{-5} M.

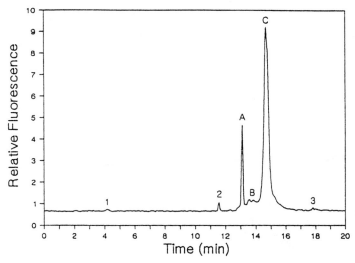

Figure 6.10 Laser-induced fluorescence detection of a hemolysate of human erythrocytes. Conditions: capillary, 110 cm × 20 μm I.D. fused silica capillary; voltage, −29 kV; detection, laser-induced fluorescence (LIF), argon ion laser operating at 275.4 nm. Peaks: A, carbonic anhydrase; B, methemoglobin; C, hemoglobin A; 1, 2, and 3, unknown. (Reprinted from Ref. 49 with permission.)

Fluorescence detection has been applied almost exclusively to the analysis of biomolecules. Applications for cations have been reported, however, including one that used indirect fluorescence for the analysis of a mixture of five alkali and alkaline earth metals[52] within 4 min, and another that employed complexing equilibria to detect calcium, magnesium, and zinc[53], directly. Detection limits on the order of 10^{-5} M for cations have been reported.

6.5.4 Electrochemical Detectors

Conductivity detection was first used in capillary electrophoresis by Mikkers et al.[54] in 1979, when they employed a potential drop between two electrodes separated along the flow direction. However, until the late 1980s conductivity detection was rarely used in CE, owing to the difficulty of fabricating a conductivity detector with low dead volume inside a fused-silica capillary with an inside diameter of 100 μm or less.

The major difficulty in designing a conductivity detector is preventing the high separation potentials from interfering with the electrochemical process. Three possible designs for a CE conductivity detector are on-column, off-column, and end-column structures. Although on-column detection is reported most commonly, there is the problem of how to produce such structures reliably and inexpensively. The problem with end-column detection is designing the instrument without the extra zone broadening typical of end-column detection systems. Huang et al.[55] constructed an on-column detector by fixing platinum wires through diametrically opposite holes in 50- or 75-μm-I.D. fused silica capillary tubing. The 40-μm-I.D. holes were made with a computer-controlled CO_2 laser. Under a microscope, two 25-μm-O.D. platinum wires were placed in the holes exactly opposite to one another in order to minimize the potential difference between the electrodes when a high electric field is applied. Because CE typically has a voltage drop of 300 V/cm along the length of the capillary, the sensing electrodes need to be aligned carefully, and an isolation transformer must be used in measuring conductance.[55] With that instrumental design, Huang et al.[55] noted excellent resolution, owing to the very small cross-sectional area of the electrodes. There is essentially no dead volume in the detector, and the detection volume is very small.

Huang and co-workers[56,57] have used the conductivity detection system for a number of applications, including the analysis of lithium in blood serum and the quantitiative analysis of low molecular weight carboxylic acids. Peak area was linear with concentration, and detection limits for lithium were 10^{-7} M, even in a large excess of sodium. Quantitation was linear over three orders of magnitude and had a variation of less than 5%. Beckers and co-workers[58] also described the use of on-column conductivity detection, using a closed system in which the EOF was suppressed. They applied the system to confirm, experimentally, the theoretical relationship between peak area and effective mobility for fully ionized, monovalent anions. As expected, the data showed a linear relationship between peak area and migration time, passing through the origin and being independent of the nature of the ionic species.

Off-column conductivity detection is achieved by grounding the capillary prior to the sensing electrode,[59] and it is exemplified by the on-column frit structure designed by Huang and Zare.[60] The frit allows an electrical connection to be made to the capillary so that the first segment of the capillary may be used for the separation while the second segment is free of applied field, facilitating its use for electrochemical detection. The design was used for continuous sample collection, and no external dilution was caused by the collection procedure.

An end-column detection scheme was designed by Huang et al.[61] in which the detector was placed directly at the outlet of the capillary. The detector was easy to construct and did not suffer from electrical interferences caused by the applied high voltage. Under typical operating conditions the extra zone broadening was less than 25%. An improved end-column conductivity detector was also described by Huang and Zare[59] in which the sensing electrode was inserted into the end of the capillary to the end of a small hole that had been drilled using a computer-controlled CO_2 laser. With the detector, the extra zone broadening was less than 15%, and the setup was applied to the detection of 5×10^{-5} M of a group of metal cations.

Suppressed conductivity detection for CE has been reported by both Dasgupta[62] and Avdalovic[63] and co-workers for the analysis of inorganic and organic anions. Both groups claim an increase in sensitivity for anions over indirect absorption of one to two orders of magnitude. Although Dasgupta and Bao[62] did not apply their detector to the analysis of cations, both groups agreed that the capillary would need to be coated for that application. Avdalovic et al.[63] used polyethyleneimine to coat a capillary prior to separating a group of low-mobility amines by suppressed conductivity CZE. Figure 6.11 shows the analysis of 10 μM each of 16 carboxylic acids by suppressed conductivity detection.

Wallingford and Ewing[64] were the first to report amperometric detection with CE. They overcame the problem of interfacing the applied separation potential with the amperometric detector by decoupling the electrochemical detector from the separation capillary using a small, electrically conductive joint near the end of the capillary. The portion of the capillary preceding the joint serves to separate the analytes; detection is performed beyond the joint by insertion of a single carbon fiber into the capillary. The performance of the electrochemical detector was optimized by miniaturizing both the separation capillary (12.7 μm I.D.) and the electrochemical

6.5 Detection

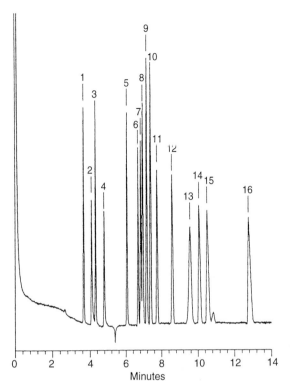

Figure 6.11 Separation of carboxylic acids (10 mM each) by suppressed conductivity capillary electrophoresis. Conditions: capillary, 60 cm × 75 μm I.D. fused silica; voltage, +24 kV; detection, suppressed conductivity using 15 mN sulfuric acid as regenerant. Peaks (ppm): 1, quinic (1.92); 2, benzoic (1.44); 3, lactic (0.90); 4, acetic (0.60); 5, phthalic (1.66); 6, formic (0.46); 7, succinic (1.18); 8, malic (1.34); 9, tartaric (1.50); 10, fumaric (1.16); 11, maleic (1.16); 12, malonic (1.04); 13, citric (1.92); 14, isocitric (1.92); 15, cis-aconitic (1.74); 16, oxalic (0.90). (Reprinted from Ref. 63 with permission.)

detector. Detection limits of about 10^{-9} M for catechols were obtained. Indirect off-column amperometric detection has also been reported, using a carbon fiber electrode placed in the end of the capillary. 3,4-Dihydroxybenzylamine was used as the electrolyte to provide the background signal, and the system was applied to the detection of amino acids and peptides.

Figure 6.12 shows the detection of carbohydrates by amperometric detection at a constant potential (0.6 V versus Ag/AgCl) with a cylindrical copper wire electrode[65] (25 μm in diameter). Because the pK values of most sugars are in the vicinity of 12–13, they are ionized at high pH and separated by CZE under such conditions, without prior derivatization or complexation procedures.

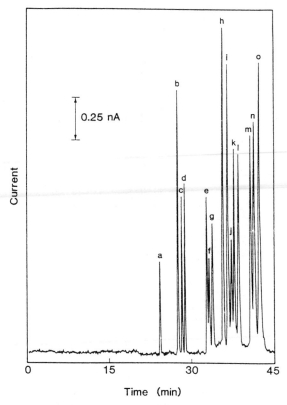

Figure 6.12 Amperometric detection of a mixture of 15 different carbohydrates (80–150 μM). Conditions: electrolyte, 100 mM sodium hydroxide; capillary, 73 cm × 50 μm I.D. fused silica; injection, gravity, 10 cm for 10 sec; voltage, 11 kV. Peaks: a, trehalose; b, stachyose; c, raffinose; d, sucrose; e, lactose; f, lactulose; g, cellobiose; h, galactose; i, glucose; j, rhamnose; k, mannose; l, fructose; m, xylose; n, talose; o, ribose. (Reprinted from Ref. 65 with permission.)

6.5.5 Mass Spectrometry Detectors

The use of mass spectrometry (MS) as a detection system is inevitable in the evolution of any separation method, especially CE where the liquid flow rate (~1 ml/min) is compatible with conventional mass spectrometers. The combination of a high-efficiency liquid-phase separation technique, such as capillary electrophoresis, with MS detection provides a powerful system for the analysis of complex mixtures. Analyte sensitivity and the mass spectrum obtained depend on the electrospray ionization (ESI) voltage, ion-focusing parameters, and buffer composition. In general, the greatest sensitivity is obtained by employing conditions that facilitate desolvation and minimize cluster formation.[47] Three ways of interfacing for CE–MS have been described: electrospray, aerospray or ion spray, and continuous-flow fast atom bombardment. All three are modifications of LC–MS interfaces.

(i) Capillary Electrophoresis–Electrospray Ionization Mass Spectrometry

The first approach to interfacing CE with MS was reported by Smith et al.[48] when they incorporated the electrospray ionization (ESI) technique introduced by Dole et al.[66] This development was based on the recognition that it is not necessary for the detection end of the CE capillary to be immersed in the buffer reservoir as conventionally practiced, as long as it is biased negative of the cathode potential (assuming a cathodic detector end). The ESI was created directly at the terminus of the CE capillary, avoiding any postcolumn region that would contribute to extracolumn band spread or analyte adsorption. A quadrupole mass filter was combined with ESI to produce the first on-line MS detection with capillary electrophoresis.

Operation of both CE and ESI requires an uninterrupted electrical contact for the electroosmotically eluting liquid at the capillary terminus. In the earlier versions of the CE–ESI interface, the electrical contact necessary to establish both the ESI and CE electric fields was implemented using an electrodeposited metal contact at the end of the CE capillary. Electrospray ionization was carried out at atmospheric pressure by biasing the ESI focusing ring with respect to the metallized column outlet. Hot nitrogen flowed across the electrospray to drive off solvent before the electrosprayed ions entered the mass analyzer. With this configuration, detection limits on the order of 14–17 fmol injected have been reported for quaternary ammonium salts.[67]

An improved electrospray ionization interface for CE–MS was developed in 1988 by Smith et al.[68] in which the metal contact at the CE terminus was replaced with a thin sheath of flowing liquid. The electrical contact was achieved through a conductive liquid sheath, typically methanol, acetonitrile, acetone, or isopropanol. One advantage of the configuration over the previous one was a qualitative improvement in ESI stability. In addition, the CE capillary could be easily replaced, as metal no longer had to be deposited at the end of the capillary. The flowing liquid sheath electrode interface allows the composition of the electrosprayed liquid to be controlled independently of the CE electrolyte, which is a desirable property because the high percentage aqueous and high ionic strength electrolytes commonly employed in CE are not well tolerated by ESI208. The sheath flow electrode interface did not degrade CE separations and greatly extended the utility of CE–MS by allowing operation over an essentially unlimited range of flow rates and buffer compositions.[68] That CE–MS configuration has been applied to the analysis of nucleotide mono-, di-, and triphosphates,[69] nucleotide coenzymes,[69] polypeptides,[70] and proteins,[70] with detection limits for myoglobin and cytochrome c of approximately 100 fmol.

(ii) Capillary Electrophoresis–Ion Spray Mass Spectrometry

Lee et al.[71] reported an alternative liquid junction for coupling CE to an ion spray chromatograph–mass spectrometer interface on a commercially available, atmospheric pressure ionization triple-quadrupole (tandem) mass spectrometer. The authors claimed that electrospray nebulization is improved by pneumatic assistance. The ion spray (pneumatically assisted electrospray) LC–MS interface offers all the benefits of electrospray ionization with the additional advantages of accommodating a wider liquid flow range, higher percentage of water in the mobile phase, and higher buffer concentrations and improving ion current stability.[71] Detection limits on the order of high parts per billion to low parts per million were reported. On-line CE–ion spray tandem MS has also been reported for the analysis of small peptides,[70] with low femtomole detection limits under selected ion monitoring conditions.

(iii) Capillary Electrophoresis–Continuous-Flow Fast Atom Bombardment Mass Spectrometry

The combination of CE with continuous-flow fast atom bombardment (CF-FAB) MS requires the use of an interface because of the incompatibility of the CF-FAB process and CE for liquid flow. The CF-FAB source requires a solvent, usually water/glycerol (95:5 v/v), which is maintained at a steady flow rate of 2–15 ml/min. Flow rate in CE, on the other hand, usually does not exceed 1 nl/min.[72] Flowing FAB interfaces have the attraction over electrospray ionization interfaces of compatibility with conventional mass spectrometers. However, limitations include the problem of introducing FAB matrix substances such as glycerol, the problem of minimizing any pressure gradient across the capillary while avoiding long transfer lines and the chemical noise and sensitivity constraints inherent in the ionization method.[70]

A coaxial CF-FAB interface was applied to the coupling of CE with tandem MS. A pair of coaxial fused silica capillary columns were used to deliver, independently, the microcolumn effluent and the FAB matrix directly to the FAB probe tip face. The advantages of the system are that the composition and flow rates of the two liquid streams can be independently optimized, the FAB matrix does not affect the microcolumn separation process, and peak broadening is minimized because the two streams do not mix until they reach the tip of the FAB probe, where ion desorption occurs.

An interface has been connected to a high-resolution double-focusing mass spectrometer applied to the analysis of mixtures of both chemically synthesized peptides and peptides obtained from proteolytic digests of proteins. Detection limits of about 35 fmol were reported. Reinhold et al.[73] described the on-line coupling of CE and MS using a CF-FAB system in combination with a liquid junction that had been designed as a coupling device for postcapillary derivatization in laser-induced fluorescence detection. The liquid junction, which is a low-dead-volume T piece between the separation capillary and the transfer capillary, results in some loss of resolution, but detection limits in the low femtomole range were reported for the drug dextromethorphan.

6.6 Fraction Collection

The major difficulty with developing an interface that couples CE with fraction collection is the fact that in CE only a limited amount of sample can be introduced into the capillary. This means that little analyte is available for downstream collection and analysis. Several approaches to the problem have been devised. One approach is to collect the fractions into vials. The problem with that approach is the possibility of significant dilution of the collected analyte, which could reduce the analyte concentration below detection limits. A second problem is the remixing of the separated species in the outlet reservoir.

Another approach to fraction collection is the use of an on-column frit structure or capillary fracture that depends on the electroosmotic flow to deposit the eluent in a continuous manner on a moving surface. Although this approach circumvents the dilution problem, the collection structures are complex and can result in the loss of some of the analyte. One commercially available fraction collection device couples CE with membrane fraction collection, without the need for frits or capillary fractures. The outlet vial holder can be removed and replaced with a wetted circular polyvinylidene difluoride (PVDF) disk, which enables the collection of eluted analytes and subsequent manipulations such as immunoblotting and microsequencing. Figure 6.13 shows a schematic diagram of the CE membrane fraction collector interface.[74]

6.7 Sample Preparation

Sample preparation in CE is similar to sample preparation in HPLC (Section 3.8). It is usually performed off-line in CE, because commercial instrumentation does not allow for on-line sample preparation. The exception is sample preconcentration, which is readily performed on-line. Sample preparation includes any manipulation of the sample, such as weighing, filtration, dilution, or derivatization, prior to introduction into the CE system.

Successful results in CE often depend on the use of appropriate sample handling techniques. This requires knowing the nature of the analytes and their behavior under various conditions. For example, the main difference between the analysis of inorganic species in CE and HPLC is the size or

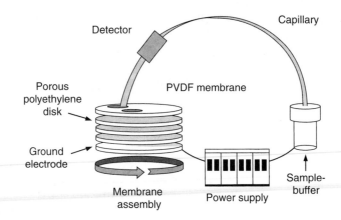

Figure 6.13 Membrane fraction collector interface for CE. (Reprinted from Ref. 74 with permission.)

volume of the sample. In CE, sample volumes tend to be much smaller than in HPLC, so it is important to ensure that all sample preparation devices are free from contamination. It is often better in CE to handle the sample as little as possible, and "clean" samples for ion analysis are often injected without filtration in order to avoid the introduction of ionic contaminants, such as chloride or sulfate. Sample preparation is discussed in more detail in Chapter 3.

For the analysis of denatured proteins by CE, it is important to include specific denaturants, such as ultrapure, electrophoresis-grade urea (4 to 8 M), at sufficient concentrations in the sample to ensure complete denaturation. Occasionally, it may be necessary to add urea to the run buffer, but the concentration should be as low as possible, as urea has a significant absorbance in the low-UV range. If a denaturant is not included, because unfolding may be incomplete and conformers or aggregates can be formed, the analytes can appear during electrophoresis as low, broad conformer peaks that comigrate with native protein peaks. For the analysis of native proteins, sulfhydryl reagents, which are added to prevent oxidation, often need to be replenished. Membrane proteins require surfactants to stay in solution, and these should not be removed. The pH of the sample should be at or below that of the running buffer, to ensure quantitative results. Thus, injection of proteins from water is not recommended, as no buffering capacity exists and the pH is uncertain.

6.8 Troubleshooting

Troubleshooting in CE is not as well defined as it is in HPLC, and, in general, the instrumentation is not readily serviced by the user. Nevertheless, there are several steps the user may take in order to identify and rectify problems that can interfere with analyses (Table 6.2). Preventive maintenance is always the best approach, and buffers and samples should be filtered unless contamination from the filters is an issue; with the narrow capillaries used in CE, blockage with particulate matter is a common problem.

An important step for ensuring the reproducibility of a method in CE is the conditioning of the capillary. For uncoated fused silica capillaries, it is advisable to precondition the capillary by washing it with a base solution such as sodium hydroxide before the first run. This will normalize the capillary interior from capillary to capillary and reduce changes in electroosmotic flow during the "break-in" period. For samples containing species that interact strongly with the capillary wall, it is usually necessary to wash the capillary with hydroxide between runs, to prevent changes in the velocity of the electroosmotic flow. For samples containing small neutral molecules (MECC) or inorganic or low molecular weight organic ions (CZE), between-run rinses with hydroxide are not necessary. When using coated capillaries or gel-filled capillaries, it is important to follow the manufacturer's recommendations for capillary treatment.

Table 6.2 CE Problems, Probable Causes, and Remedies

Problem No. 1: No peaks/very small peaks		
Problem	Probable cause	Remedy/Comments
Normal / Problem / Problem	1. Capillary not aligned in detector	1. Align capillary.
	2. Plug/bubble in capillary	2. Purge with a syringe.
	3. No voltage	3. Confirm voltage setting. Ensure capillary ends are immersed in buffer.
	4. No sample injected	4. Confirm sample in vial. Be sure capillary extends into sample. Confirm injection time.
	5. Detector lamp off/dead	5. Turn lamp on. Replace lamp.
	6. Incorrect detector wavelength	6. Confirm wavelength setting. Confirm wavelength accuracy.
	7. Wrong buffer	7. Confirm buffer composition.

(*continues*)

Table 6.2 *continued*

Problem	Probable cause	Remedy/Comments
Problem No. 2: No Current		
(Normal / Problem traces)	1. Plug/bubble in capillary	1. Purge capillary with a syringe. Trim capillary ends. Replace capillary.
	2. Broken capillary	2. Replace capillary.
	3. Safety interlock not closed	3. Close interlock.
	4. Wrong buffer	4. Confirm buffer composition.
Problem No. 3: Poor sensitivity		
(Normal / Problem traces)	1. Incorrect injection volume	1. Confirm injection time, voltage, pressure, etc.
	2. Aging detector lamp	2. Replace lamp.
	3. Incorrect detector wavelength	3. Confirm wavelength setting.
Problem No. 4: Current fluctuations		
(Normal / Problem traces)	1. Bubbles in capillary	1. Flush capillary with a syringe.
	2. Temperature fluctuations	2. Monitor capillary/room temperature.

(*continues*)

6.8 Troubleshooting

Table 6.2 *continued*

Problem No. 5: Variable migration times		
Problem	Probable cause	Remedy/Comments
Normal / Problem / Problem	1. Temperature fluctuations 2. Ion depletion of buffers 3. Ionic strength differences between samples 4. Sample-wall interactions 5. Capillary not washed	1. Measure capillary/room temperature. 2. Replace buffers. 3. Use reference compounds. 4. Wash capillary thoroughly. 5. Inject electroosmotic flow marker.
Problem No. 6: Baseline Noise		
Normal / Problem / Problem	1. Aging detector lamp 2. Wrong detector wavelength 3. Mismatched buffers 4. Capillary not conditioned 5. Particles in buffer 6. Bubbles in capillary, due to pinholes or broken capillary	1. Replace lamp. 2. Confirm wavelength setting. 3. Confirm buffer composition. 4. Rinse with base, followed by water and buffer. 5. Filter buffer. 6. Replace capillary.

(continues)

Table 6.2 *continued*

Problem No. 7: Broad Peaks		
Problem	Probable cause	Remedy/Comments
Normal / Problem	1. Sample-wall interaction	1. Wash capillary well.
	2. Too large an injection	2. Reduce injection time.
	3. Joule heating	3. Reduce voltage. Decrease buffer concentration.
	4. Buffer siphoning	4. Equalize buffer reservoir levels.
	5. Ionic strength of sample diluent too high	5. Dilute sample in water. Remove salt from sample. Increase concentration of buffer.

6.9 Summary of Major Concepts

1. The main components of a CE instrument are

power supply,
injection system,
capillary,
detector, and
buffer vials.

2. The high-voltage DC power supply provides the electric field necessary to establish the electroosmotic flow of the bulk solution and also the electromigration of the charged analytes. Most power supplies provide voltages from -30 kV to $+30$ kV.

3. Sample is introduced into the capillary by hydrostatic injection (gravity, pressure, or vacuum) or by electromigration injection (voltage).

4. The most common capillaries used in CZE or MECC for the analysis of small molecules are untreated fused silica capillaries. The analysis of large proteins by CZE and applications for cIEF are generally performed in coated capillaries. The CGE technique is performed in gel-filled capillaries or capillaries filled with polymer networks.

5. Ultraviolet absorbance detection is the most prevalent type of detection in CE, and UV detectors operate in both the direct and indirect modes. Laser-induced fluorescence detection is often used for high-sensitivity work. Conductivity detection, suppressed conductivity detection, and mass spec-

trometry detection have started to appear in commercial instruments, and applications in the literature should become more prevalent as a consequence.

6. The main problem interfacing a fraction collector with CE is the small sample volumes injected in CE. Many commercial instruments provide a means for sample collection.

References

1. Albin, M., Grossman, P. D., and Moring, S. E., *Anal. Chem.* **65**, 489A (1993).
2. Chien, R.-L., and Burgi, D. S., *Anal. Chem.* **64**, 489A (1992).
3. Maystre, F., and Bruno, A. E., *Anal. Chem.* **64**, 2885 (1992).
4. Bondoux, G., Jandik, P., and Jones, P. R., *J. Chromatogr.* **602**, 79 (1992).
5. Jandik, P., and Jones, W. R., *J. Chromatogr.* **546**, 431 (1991).
6. Wildman, W. J., Jackson, P. E., Jones, W. R., and Alden, P. G. *J. Chromatogr.* **546**, 459 (1991).
7. Harrold, M., Wojtusik, M. J., Riviello, J., and Henson, P., *J. Chromatogr.* **640**, 463 (1993).
8. Jackson, P. E., and Haddad P. R., *J. Chromatogr.* **640**, 486 (1993).
9. Weston, A., Brown, P. R., Heckenberg, A. L., Jandik, P., and Jones, W. R., *J. Chromatogr.* **602**, 249 (1992).
10. Weston, A., Brown, P. R., Jandik, P., Heckenberg, A. L., and Jones, W. R., *J. Chromatogr.* **608**, 395 (1992).
11. Chen, M., and Cassidy, R. M., *J. Chromatogr.* **640**, 425 (1993).
12. Shi, Y., and Fritz, J. S., *J. Chromatogr.* **640**, 473 (1993).
13. Riviello, J. M., and Harrold, M. P., *J. Chromatogr.* **652**, 385 (1993).
14. "FMOC-Derivatized Amino Acids by Capillary Electrophoresis," Application Note 134, LPN 034822 5M 11/92. Dionex Corporation, Sunnyvale, California, 1992.
15. Stover, F. S., Haymore, B. L., and McBeath, R. J., *J. Chromatogr.* **470**, 241 (1989).
16. Bullock, J. A., and Yuan, L.-C., *J. Microcolumn Sep.* **3**, 241 (1991).
17. Lauer, H. H., and McManigill, D., *Anal. Chem.* **58**, 166 (1986).
18. Hjerten, S., *Chromatogr. Rev.* **9**, 122 (1967).
19. Cobb, K. A., Dolnik, V., and Novotny, M., *Anal. Chem.* **63**, 2478 (1990).
20. Swedberg, S. A., *Anal. Biochem.* **185**, 51 (1990).
21. Bruin, G. J. M., Chang, J. P., Kuhlman, R. H., Zegers, K., Kraak, J. C., and Poppe, H., *J. Chromatogr.* **471**, 429 (1989).
22. Towns, J. K., and Regnier, F., *J. Chromatogr.* **516**, 69 (1990).
23. Hjerten, S., and Kubo, K., *Electrophoresis* (*Weinheim*) **14**, 390 (1993).
24. Huang, M., and Lee, M. L., *J. Microcolumn Sep.* **4**, 491 (1992).
25. Lee, C. S., Blanchard, W. C., and Wu, C.-T., *Anal. Chem.* **62**, 1550 (1990).
26. Hjerten, S., *J. Chromatogr.* **347**, 191 (1985).
27. Bruin, G. J. M., Huiden, R., Kraak, J., and Poppe, H., *J. Chromatogr.* **480**, 339 (1989).
28. Huang, M., Vorkink, W. P., and Lee, M. L., *J. Microcolumn Sep.* **4**, 135 (1992).

29. Malik, A., Zhao, Z., and Lee, M. L., *J. Microcolumn Sep.* **5**, 119 (1993).
30. Bruin, G., Tock, P., Kraak, J., and Poppe, H., *J. Chromatogr.* **517**, 557 (1990).
31. Dougherty, A. M., Wolley, C. L., Williams, D.,L., Swaile, D. F., Cole, R. D., and Sepaniak, M. J., *J. Liq. Chromatogr.* **14**, 907 (1991).
32. Li, S. F. Y., "Capillary Electrophoresis." *J. Chromatogr. Libr.*, **52**, Chapter 4 (1992).
33. Chervet, J.-P., Van Soest, R. E. J., and Ursem, M., *J. Chromatogr.* **543**, 439 (1991).
34. "Model 270A-HT High throughput Capillary Electrophoresis System," Applied Biosystems, No. 127203, Order No. L-1663, Perkin Elmer Corporation, Foster City, CA, 1993.
35. Gordon, G. B., *U.S. Patent 5,061,361*, Oct. 29, 1991.
36. "HP3D Capillary Electrophoresis System—Technical Description," Publ. 12-5091-6542E, Hewlett-Packard Company, Foster City, CA, 1993.
37. Li, S. F. Y., in "Capillary Electrophoresis: Principles, Practice and Applications," Chapter 3. Elsevier, Amsterdam, 1992.
38. Jandik, P., and Bonn, G., in "Capillary Electrophoresis of Small Molecules and Ions," Chapter 3. VCH, Publ New York, 1993.
39. Grossman, P. D., and Colburn, J. C., in "Capillary Electrophoresis: Theory and Practice." Academic Press, San Diego, 1992.
40. Vindevogel, E. D., Schuddinck, G., Dewaele, C., and Verzele, M., *HRC & CC, J. High Resolut. Chromatogr. Chromatogr. Commun.* **11**, 317 (1988).
41. Foret, F., Deml, M., Kahle, V., and Bocek, P., *Electrophoresis (Weinheim)* **7**, 430 (1986).
42. Brownlee, R. G., and Compton, S. W., *Am. Biotechnol. Lab.* **6**, 10 (1988).
43. Kobayashi, S., Ueda, T., and Kikumoto, M., *J. Chromatogr.* **480**, 179 (1989).
44. Yeo, S. K., Lee, H. K., and Li, S. F. Y., *J. Chromatogr.* **585**, 133 (1991).
45. Hjerten, S., Elenbring, K., Kilar, F., Liao, J.-L., Chen, A. J., Siebert, C. J., and Zhu, M.-D., *J. Chromatogr.* **403**, 47 (1987).
46. Foret, F., Fanali, S., Nardi, A., and Bocek, P., *Electrophoresis (Weinheim)* **11**, 70 (1990).
47. Wallingford, R.A., and Ewing, A. G., *Adv. Chromatogr.* **29**, 1 (1989).
48. Smith, R. D., Olivares, J. A., Nguyen, N. T., and Udseth, H. R., *Anal. Chem.* **60**, 436 (1988).
49. Lee, T. T., and Yeung, E. S., *Anal. Chem.* **64**, 3045 (1992).
50. Green, J. S., and Jorgenson, J. W., *HRC & CC, J. High Resolut. Chromatogr. Chromatogr. Commun.* **7**, 529 (1984).
51. Kuhr, W., and Yeung, E. S., *Anal. Chem.* **60**, 1832 (1988).
52. Gross, L., and Yeung, E. S., *Anal. Chem.* **62**, 427 (1990).
53. Kaniansky, D., Rajec, P., Svec, A., Marak, J., Koval, M., Lucka, M., Francko, S., and Sabanos, G., *J. Radioanal. Nucl. Chem.* **129**, 305 (1989).
54. Mikkers, F. E. P., Everaerts, F. M., and Verheggen, T. P. E. M., *J. Chromatogr.* **169**, 11 (1979).
55. Huang, X., Pang, T.-K., Gordon, M. J., and Zare, R. N., *Anal. Chem.* **59**, 2747 (1987).
56. Huang, X., Gordon, M. J., and Zare, R. N., *J. Chromatogr.* **425**, 385 (1988).
57. Huang, X., Gordon, M. J., and Zare, R. N., *J. Chromatogr.* **480**, 285 (1989).

References

58. Ackermans, M. T., Everaerts, F. M., and Beckers, J. L., *J. Chromatogr.* **549**, 345 (1991).
59. Huang, X., and Zare, R. N., *Anal. Chem.* **63**, 2193 (1991).
60. Huang, X., and Zare, R. N., *Anal. Chem.* **62**, 443 (1990).
61. Huang, X., Zare, R. N., Sloss, S., and Ewing, A. G., *Anal. Chem.* **63**, 192 (1991).
62. Dasgupta, P. K., and Bao, L., *Anal. Chem.* **65**, 1003 (1993).
63. Avdalovic, N., Pohl, C. A., Rocklin, R. D., and Stillian, J. R., *Anal. Chem.* **65**, 1470 (1993).
64. Wallingford, R. A., and Ewing, A. G., *Anal. Chem.* **59**, 1762 (1987).
65. Colon, L. A., Dadoo, R., and Zare, R. N., *Anal. Chem.* **65**, 476 (1993)
66. Dole, M., Mack, L. L., Hines, R. L., Mobley, R. C., Ferguson, L. D. and Alice, M. B., *J. Chem. Phys.* **49**, 2240 (1968).
67. Seng, H. P., *Text. Prax. Int.* **39**, 795 (1984).
68. Smith, R. D., Barinaga, C. J., and Udseth, H. R., *Anal. Chem.* **60**, 1948 (1988).
69. Olivares, J. A., Nguyen, N. T., Yonker, C. R., and Smith, R. D., *Anal. Chem.* **59**, 1230 (1987).
70. Smith, R. D., Loo, J. A., Barinaga, C. J., Edmonds, C. G., and Udseth, H. R., *J. Chromatogr.* **480**, 211 (1989).
71. Lee, E. D., Muck, W., Henion, J. D., and Covey, T. R., *Biomed. Environ. Mass Spectrom.* **18**, 844 (1989).
72. Caprioli, R. M., Moore, W. T., Martin, M., DaGue, B. B., Wilson, K., and Moring, S., *J. Chromatogr.* **480**, 247 (1989).
73. Reinhold, N. J., Niessen, W. M. A., Tjaden, U. R., Gramberg, L. G., Verheij, E. R., and van der Greef, J., *Rapid Commun. Mass Spectrom.* **3**, 348 (1989).
74. Warren, W. J., Cheng, Y-F., and Fuchs, M., *LC-GC* **12**, 22 (1994).

CHAPTER 7

Data Manipulation

7.1 Introduction

There are two purposes for using HPLC or CE: determining the nature of the analytes in a sample and determining the concentration of each analyte. The former purpose is known as qualitative analysis, and the latter is termed quantitative analysis. The sample passes through the instrument and generates a signal that is recorded by the data station or strip-chart recorder. The signal must then be converted into qualitative or quantitative information. Data manipulation is the final step of the analysis.

7.2 Identification of Peaks

A feature of HPLC and CE is that, for a defined system with invariant conditions, an analyte will have a constant retention volume and thus may be identified on the basis of retention time. Given the numerous compounds in existence, however, it is not possible to assign, unequivocally, an identity to any peak on the basis of retention times alone, unless the individual components of a sample mixture are known prior to analysis.

There are a variety of techniques currently in use to aid in the identification of sample components. Most techniques, such as "spiking" and the enzyme peak shift method, are used to confirm the identities of components thought to be in the sample. If, however, the identity of the sample is truly unknown, a combination of techniques is needed to provide a unique "fingerprint."

7.2.1 Use of Retention Data

As a first approximation, the most commonly used method of peak identification is that of matching the retention times, or occasionally capacity factors, of the sample components with those of standard reference

7.2 Identification of Peaks

compounds. To minimize the effects of the sample matrix on the retention times the sample should be analyzed under conditions identical to, or similar to, those used for the standards. Figure 7.1 shows the reversed-phase separation of two bile acid samples run under identical conditions.[1] One sample was a standard containing 15 free and conjugated bile acids (Fig. 7.1a), whereas the other was a serum sample from a healthy volunteer (Fig. 7.1b). By comparing the retention times of the peaks in the two chromatograms it is possible to assign, tentatively, the identities of most

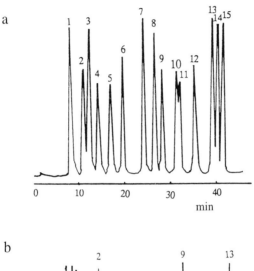

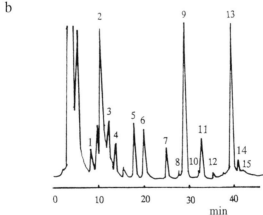

Figure 7.1 Use of standards (a) for the identification of unknowns (b) by retention times. Peaks: 1, ursodeoxycholic acid; 2, glycine-ursodeoxycholic acid; 3, cholic acid; 4, taurine-ursodeoxycholic acid; 5, glycine-cholic acid; 6, taurine-cholic acid; 7, chenodeoxycholic acid; 8, deoxycholic acid; 9, glycine-chenodeoxycholic acid; 10, glycine-deoxycholic acid; 11, taurine-chenodeoxycholic acid; 12, taurine-deoxycholic acid; 13, lithocholic acid; 14, glycine-lithocholic acid; 15, taurine-lithocholic acid. (Reprinted from Ref. 1 with permission.)

of the peaks in the serum sample. However, because there are so many variables in a chromatographic system, and because there may be unexpected compounds with the same retention times, retention times can only be used for tentative or preliminary identification.

In most data systems, a given peak in a sample mixture is identified if its retention time falls within a user-defined retention time window. The retention time window is typically centered about the retention time of that peak in a standard solution (i.e., $\pm 5\%$). However, the size of the window may be altered depending on the number of other components in the sample that may fall within a given window and the reproducibility of the method. The more reproducible the retention times, the smaller are the windows that may be defined.

The relative retention of a compound on two different systems can also be used for tentative peak identification. This method is based on the premise that the possibility of different compounds showing identical behavior under different conditions is relatively small. Thus, use of different detectors, for example, fluorescence and UV detection, or different techniques, such as chromatography and electrophoresis, will help to confirm the identities of peaks in a sample.

7.2.2 Standard Addition or Spiking

In the standard addition method, the sample is analyzed first. A small, known amount of a known compound, thought to be in the sample, is then added to the sample, and the mixture is reanalyzed. If a quantitative increase is obtained in the peak of the compound corresponding to the standard, the compound representing the peak is tentatively identified. The technique of spiking is applicable provided that the sample is known to contain certain components; appropriate standards can then be added to enable each peak to be identified. If the identities of the components in the sample are totally unknown, the spiking technique is inappropriate, and additional analytical or comparative techniques will be required.

7.2.3 Internal Standard

The internal standard method of qualitative analysis is used to help identify components in a mixture when the reproducibility of retention times of the components in the sample versus the standard mixture is inadequate. The change in retention times from the standard mixture to the sample mixture is usually due to sample matrix effects that are difficult to mimic in the standard mixture. Thus, a small amount of a compound that is known to be absent from the sample is added to both the standard mixture and the sample prior to analysis.

If the retention times of the peaks do move in the sample relative to the standard, there are two methods to identify the component of interest depending on the number of peaks in the sample. If there are only a few well-resolved peaks present, and if, for example, the retention time of the internal standard increases in the sample relative to the standard, the difference in retention times of the internal standard in the standard and the sample is calculated. The difference is then subtracted from the retention times of each of the other peaks before comparing the chromatogram to the standard chromatogram. Figure 7.2 shows the use of an internal standard to help identify the components in a plasma sample.[2] The retention time of the internal standard in the sample was 1.5 min slower than in the standard. When 1.5 min was subtracted from the retention times of each of the other peaks in the sample chromatogram, the two major peaks could be assigned, tentatively to phenobarbital (4.13 min) and carbamazepine (6.62 min).

If there are many peaks in the sample, or if the peaks are eluted in close proximity to one another, it may be necessary to determine peak identities on the basis of the selectivity, α, as defined in Chapter 1 (Section 1.3.2), where t_1 refers to the retention time of the internal standard.[3] The advantage of the selectivity approach is that the relative retention ratio provides more reproducible numbers.[4]

7.2.4 Isotopic Labeling

A variation of the use of standards is the method in which predetermined quantities of a standard radioactive compound are added to the solution. The fractions are collected, and, by plotting the counts of the fractions, the peak of interest can be identified. This method is especially useful in following cell metabolism of purine and pyrimidine analogs. A plot of the nucleotides in a cell extract of schistosomes containing ^{14}C-labeled adenine and guanine nucleotides is shown in Figure 7.3.[5]

7.2.5 Enzyme Peak Shift

A method of great value in biochemistry for the verification of peak identities is the enzymatic peak shift technique.[5] The approach utilizes the specificity of enzyme reactions with a nucleotide or class of nucleotides. The technique is especially useful in the characterization of nucleotides of cell extracts, because not only is the identity of the reactant verified, but so is the identity of the product formed. With the enzyme peak shift method, one aliquot of the sample is analyzed while a second aliquot is incubated with an excess of an enzyme that catalyzes a specific reaction involving the compound of interest. After the enzyme is deactivated, the second aliquot is chromatographed.

Identification is based on the disappearance of a reactant peak or the

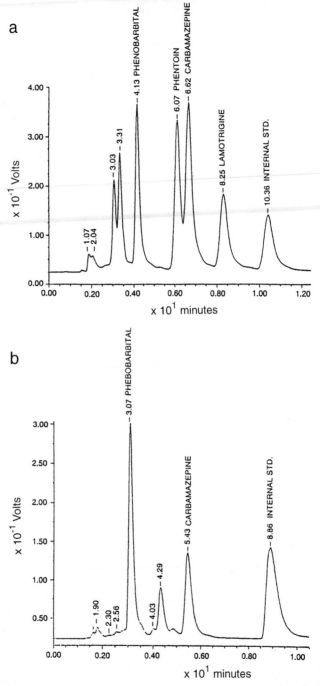

Figure 7.2 Identification of unknowns by incorporation of an internal standard. (a) Plasma spiked with standards to determine retention times under the experimental conditions, (b) extracted plasma sample from a patient receiving carbamazepine and phenobarbital. (Adapted from Ref. 2 with permission.)

7.2 Identification of Peaks

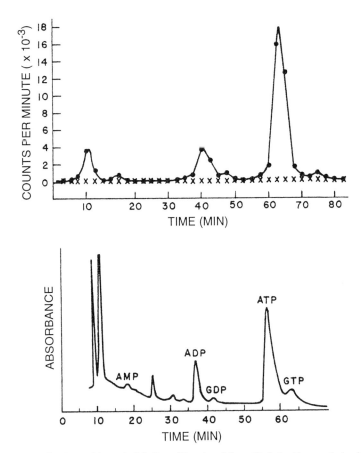

Figure 7.3 Use of isotopic labeling. (Reprinted from Ref. 5 with permission.)

appearance of a peak corresponding to the retention time of a known product. An example of the enzymatic peak shift method is the analysis of cytidine in serum samples. If an aliquot of the serum sample is incubated with the enzyme cytidine deaminase, the cytidine in that aliquot is converted to uridine.[6] When the chromatogram of the original serum sample is then compared with the chromatogram of the aliquot incubated with cytidine deaminase, the cytidine peak disappears and the uridine peak in the incubated aliquot increases in size.

Enzymatic peak shift reactions are also useful for clarifying or "unmasking" a chromatogram. If one nucleotide is present in a large quantity, as for ATP in human erythrocytes, for example, it may hide the presence of a small quantity of another compound that has a similar retention time. Therefore, if an enzyme is added that converts the large peak into a product peak with a different retention time, it becomes possible to show if hidden peaks are present.

7.2.6 Use of Ultraviolet and Mass Spectrometry Libraries

Photodiode array (PDA) detectors are becoming more popular as their ability to use both retention times and UV spectra to aid in peak identification becomes more widely recognized. With most PDA software packages, libraries of standard compounds can be created by the analyst. When samples are run, the spectra and/or the run times of the peaks in the sample can be compared with those stored in the libraries. Because the shapes of UV spectra are dependent on the experimental conditions under which they are obtained, however, it is unlikely that standard libraries will become commercially available.

An example of the use of PDA UV libraries to help in the identification of peaks is provided by the identification and quantification of tricyclic antidepressants.[7] Although the structures of the compounds are very similar, making separation fairly difficult, the UV spectra of most tricyclic antidepressants are fairly distinct, as shown in Figure 7.4. On the occasions where the UV spectra are similar, the retention times are sufficiently different to prevent misidentification.

Another feature of the photodiode array detector is the ability to determine peak homogeneity. To determine the homogeneity of a peak, software algorithms are used to compare the spectrum of the compound at the peak apex with spectra obtained at other points across the peak. In general, peak homogeneity translates into peak purity. However, if two symmetrical peaks are perfectly overlayed as illustrated in Figure 7.5, the peak, while perfectly homogeneous, is impure.

An example of the use of photodiode array detectors for determining peak homogeneity is in the analysis of β-carotene,[8] where an impurity in the all-trans isomer was positively identified as the cis isomer.[9] Since the U.S. *Federal Register* began to require listing of the presence and concentration of various vitamins on the labels of foods, there has been increased activity in the measurement of these vitamins. β-Carotene is one such vitamin, and research links the presence of β-carotene to a reduction in the occurrence of some degenerative deseases such as cardiovascular diseases and cancer.

As with PDA detection and the construction of UV libraries, computerized data systems are used for data collection and processing with mass spectral detection. Unlike UV spectra, however, mass spectra are much less dependent on experimental conditions. Thus, it is possible to purchase commercial libraries of mass spectra for spectral matching. Mass spectral data systems range from systems that record the spectra and produce conventional relative abundance bar charts for subsequent analysis to those

7.2 Identification of Peaks

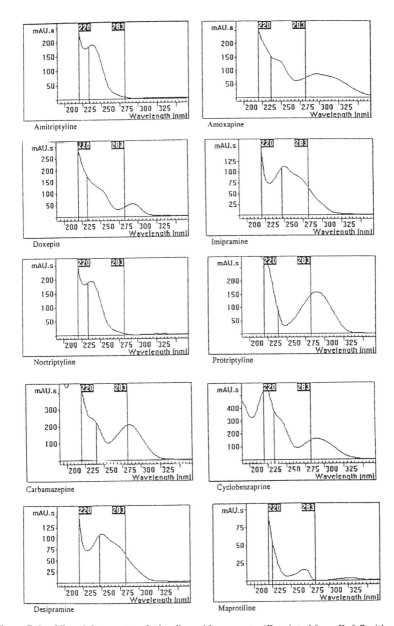

Figure 7.4 Ultraviolet spectra of tricyclic antidepressants. (Reprinted from Ref. 7 with permission.)

Figure 7.5 Coelution of two peaks illustrating perfect homogeneity.

that not only provide the bar charts but also analyze the data and compare the information with mass spectrometry libraries.

7.3 Quantitation

The other purpose for peak measurement is determining solute quantity. Although both HPLC and CE can be highly accurate analytical techniques, the quantitative results depend on the complete process of analysis, from preparation of the sample, through the separation process and detection of the signal, to interpretation of the results. There are three main sources of error in quantitation using chromatography, and ways of minimizing the first two were discussed previously. These sources of error are (1) sampling technique and sample introduction, (2) design of the instrument, and (3) peak size measurement.

This chapter describes the more commonly used methods of quantitation, with brief discussions on the theory. For more information on the use of computers and quantitation, see Braithwaite and Smith.[10] An excellent account of chromatographic integration methods for peak identification, validation, and quantitation is provided in the text by Dyson.[11]

7.3.1 Integration

The integration step consists of converting the detector signal into numerical data. Because the detector signal is in the form of a chromatogram or electropherogram from a strip-chart recorder, an integrator, or a computer, it is necessary to convert the peak area or height into numbers either by manual or electronic means.

Measurement of peak height, although an approximation, can be done with good accuracy and is simpler to perform than measurement of peak area. It is often preferred for manual data reduction because it is simple and fast. In addition, less resolution of components is required for accurate calculations.[12] However, peak heights are more susceptible to variations in separation conditions, such as column temperature and aging, than are peak areas, and the results are not as accurate as those obtained by measurement of peak areas.

There are a variety of other factors that influence the accuracy of quantitative analysis. Noise, in the form of baseline disturbances and baseline drift, affects area more than it does height, as it can cause area to be lost at the tailing edges of the peaks where they are widest. Peak asymmetry and detector saturation or nonlinearity, however, have a more detrimental effect on peak height. Figure 7.6 shows a calibration curve comparing peak height measurements with peak area measurements.[13]

(i) Manual Methods

A major disadvantage of manual methods is that the accuracy of each measurement relies on the individual performing the calculations and on where the tangents or baselines are drawn, which may differ from one peak to the next within an analysis. If strip-chart recorders are used, all the peaks must be contained on the chart paper. This severely limits the dynamic range of solute composition that can be analyzed. In addition, the chart speed must be fast enough so that, for measurements based on peak width, narrow peaks are wide enough for accurate measurement.

A. Peak Height Peak height is measured from the baseline to the peak apex, as shown in Figure 7.7. When the baseline is unstable, a baseline must be interpolated from the start of the peak to its end. Peak heights should not be used for distorted peaks or shoulders.

B. Peak Area Manual methods for converting the area of the detector signal into numerical data historically have included (1) planimetry, (2) taking the product of peak height and peak width at half-height, (3) triangu-

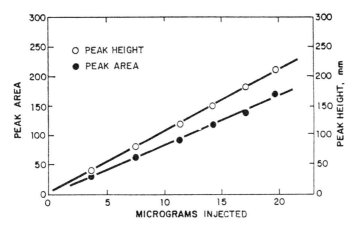

Figure 7.6 Peak height and peak area calibrations for a sample of an antioxidant. (Reprinted from Ref. 13 with permission.)

Chapter 7 Data Manipulation

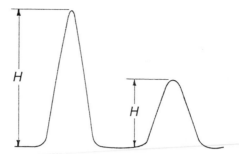

Figure 7.7 Measurement of peak height.

lation, and (4) cutting and weighing. Of these, only the second method is in common use.

The product of peak height and peak width at half height should only be used for symmetrical peaks. In the method the area is approximated by multiplying the height of the peak by the width at half-height. The width at half-height is used instead of the width at the peak base to reduce errors resulting from peak tailing or baseline irregularities. The concept is illustrated in Figure 7.8. The accuracy of the method is influenced by the width measurement, and a narrow peak can adversely affect the precision. The numerical value obtained by the approach represents only 93.9% of the theoretically true area.[14]

Two more modern and more accurate methods for manually determining peak areas are the Condal–Bosch area and Foley's equation for the manual measurement of asymmetric [exponentially modified Gaussian (EMG)] peaks. According to Figure 7.9, peak height shows the most sensitivity to peak shape, becoming progressively more inaccurate with increasing asymmetry. The method of triangulation is a more accurate method

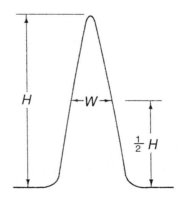

Figure 7.8 Measurement of area using peak height and peak width at half-height.

7.3 Quantitation

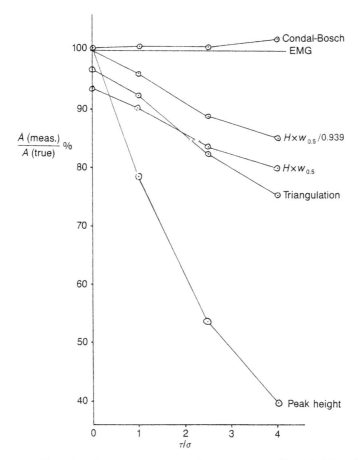

Figure 7.9 Effect of peak asymmetry on manual measurements. (Reprinted from Ref 14 with permission.)

than the measurement of width at half-height for symmetrically shaped peaks, but it becomes the less accurate method as peak asymmetry increases.

The Condal–Bosch area measurement uses the average of the peak widths at 15 and 85% of the peak height instead of the width at half-height, as shown in Figure 7.10, producing a result that is 100.4% of the true area.[11] The use of this method requires more manual measurements than other methods. After the width at 15 and 85% of the peak height has been measured, as well as the peak height, the peak area is calculated according to

$$A = H \times (w_{0.15} + w_{0.85})/2. \quad (7.1)$$

Foley's equation for the measurement of the area of a resolved, asymmetric peak requires no measurements of asymmetry and is also applicable to

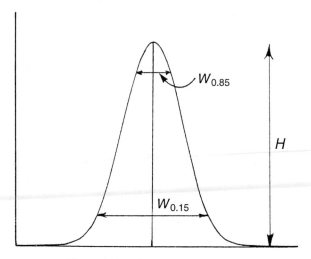

Figure 7.10 The Condal–Bosch area.

symmetrical peaks. It requires fewer measurements than the Condal–Bosch calculation and is more accurate for the measurement of symmetrical peaks than either the Condal–Bosch calculation or the more commonly used peak height times width at half-height. With the Foley approach, the width is measured at 25% of the peak height (Fig. 7.11), and the measured values are used to calculate the peak area according to Foley's equation:

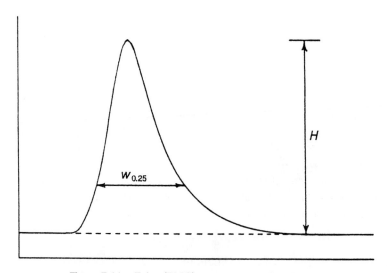

Figure 7.11 Foley (EMG) measurement of peak area.

$$A = 0.753(Hw_{0.25}). \quad (7.2)$$

The area calculated by the method is 100.3% of the true theoretical area, for a symmetrical peak.[11]

(ii) **Electronic Devices**

The most accurate measurements of peak areas are those obtained by means of electronic data reduction with integrators or computers. The vast majority of laboratories now have electronic or computer methods for determining peak areas. Two main features are required for electronic processing of chromatographic data: accurate digitization of the analog signal and software. The software is required for the detection of peaks, correction for baseline drift, calculation of peak areas, retention times, and concentrations of components in the sample, and production of the final report.[5]

The analog signal from the detector is converted into the digital binary data required by the computer interface, which includes signal buffering, amplification, and attenuation and an analog-to-digital converter (A/D converter). An A/D converter samples the analog signal at a predetermined, fixed rate set sufficiently high that the peak is described accurately. At least 10 points/sec is acceptable for most chromatography and more if a shoulder is present, otherwise detail will be lost.[5] The effect of sampling rate on resolution is shown in Figure 7.12.

Most commercially available integrators and data stations have preprogrammed software routines for standard calculations. Before the area can

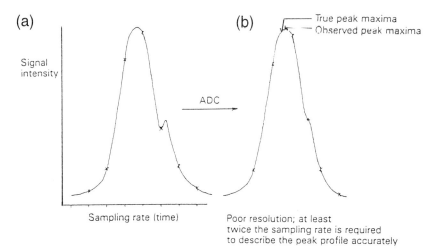

Figure 7.12 Effect of analog-to-digital converter sampling rate on resolution: (a) analog plot; (b) digital representation. (Reprinted from Ref. 10 with permission.)

be calculated, however, the software must recognize the presence of a peak, so peak detection algorithms are written to monitor a predetermined threshold above the baseline. Once the threshold is exceeded, peak monitoring occurs.

During some chromatographic separations, the baseline may drift in either a positive or a negative direction and exceed the threshold. Correcting algorithms, therefore, are incorporated into the software to adjust for baseline drift and ensure optimum peak area calculations. The slope sensitivity can be set to distinguish between the start of a peak and the start of a sloping baseline.[4] Proper measurement of peak areas requires considerable attention to these measurement options, but, when properly done, the error associated with the measurement should be under 1%. Having set the parameters in the software for detection of the peak, the peak can be integrated and the peak area determined. The area between the limits is integrated using a summation algorithm that can be based on any of a number of algorithms, such as trapezoidal integration.[10]

(iii) Sources of Error in Peak Size Measurements

There are a number of geometrical construction errors associated with manual measurements,[15,16] as seen in Figure 7.13. The most important errors are errors associated with height measurements, peak width measurements, half-height measurements, and baseline positioning. The error in measuring peak area is a combination of all four types of errors. In contrast, the error associated in measuring peak height is a combination of only peak height and baseline position measurements. Because of the requirement for fewer manual measurements, greater precision results from peak height measurements than from peak area measurements.

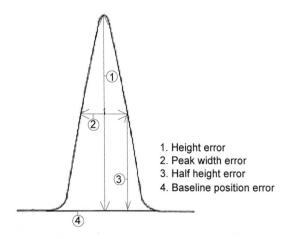

1. Height error
2. Peak width error
3. Half height error
4. Baseline position error

Figure 7.13 Errors associated with manual measurements.

7.3 Quantitation

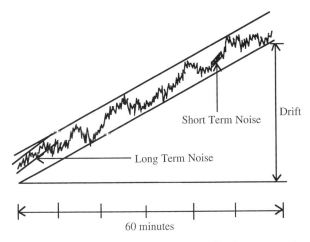

Figure 7.14 Recognized types of noise: short-term noise, long-term noise, and drift.

Inaccurate peak measurements can be caused by a variety of means, not simply by errors associated with manual measurements. Peak asymmetry, noise, baseline drift, and incompletely resolved peaks all contribute to errors in peak measurement. Unless an asymmetrical peak is measured using a method that takes peak asymmetry into account, large errors may be introduced into the measurement.

Noise can be chemical or electronic in origin and is defined as the variation in output signal that cannot be attributed to the solute passing through the cell. There are three kinds of noise recognized by the ASTM,[17] as shown in Figure 7.14. Short-term noise is defined as random variations in the detector signal with frequency greater than 1 cycle/min. Short term noise widens the trace and appears as "fuzz" on the baseline. Long-term noise is seen as variations in the detector signal with frequencies between 6 and 60 cycles/hr. Long-term noise appears as valleys and peaks that may be difficult to distinguish from real peaks at very low concentrations.

The steady movement of the baseline either up or down the scale is referred to as drift. Drift is often indicative of variations in chromatographic conditions, such as temperature or solvent programming. It can also be indicative of instrument instability owing to temperature effects on the detector.

The smallest measurable peak is defined as the smallest peak that can be distinguished, unambiguously, from the baseline noise. It is described in terms of the signal to noise ratio, S/N, which compares the height of the peak with the height of the surrounding noise, as illustrated in Figure 7.15.

According to the American Chemical Society (ACS) guidelines[18] published in 1980, the limit of detection (LOD) at $S/N = 3$ defines the smallest

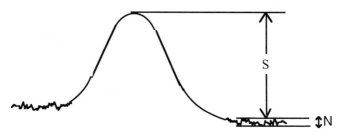

Figure 7.15 Signal-to-noise ratio.

peak that can be judged, confidently, to be a peak. The smallest detectable peak, however, is too small for accurate quantitation. Thus, the limit of quantitation (LOQ), which is the smallest peak whose area can be measured with accurate precision, is measured at $S/N = 10$.

Baseline drift creates errors in a number of ways. It can affect the true area of the peak, the retention time of the peak, and the operating range of the detector. When a peak is eluted on a baseline slope, the baseline that is drawn under the peak is usually a straight line. The true baseline, however, may often be curved, as shown in Figure 7.16; thus, area may be lost or gained by incorrect placement of the baseline.

Incompletely resolved peaks are the final obstruction to the accurate

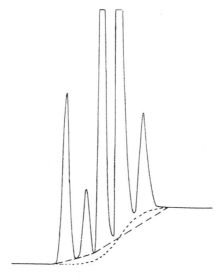

Figure 7.16 Errors associated with incorrect placement of baseline. The dotted line represents the true baseline; the dashed baseline is the one placed by the integrator. The left-hand side shows area lost by incorrect placement of the baseline; the right-hand side shows area gained by incorrect placement of the baseline. (Reprinted from Ref. 11 with permission.)

measurement of chromatographic peaks. Algorithms are written into most software packages to help determine the area under poorly resolved peaks, but it is the chromatographer who must decide which approach is the best for any given set of peaks. The most common approach, as illustrated in Figure 7.17, is the "perpendicular drop." With this approach, the valleys between merged peaks are detected by the software algorithm, which then drops a perpendicular line down to the baseline.

The perpendicular drop should only be used if (1) the peaks are symmetrical and about the same height and width, (2) the valley between the peaks is no more than 5% of the peak height, (3) the baseline is flat, and (4) noise does not obscure the accurate placement of the baseline. If the peaks are asymmetrical, a perpendicular drop will overestimate the area of the smaller peak. Under those circumstances, a tangential skim may be a better approach.

Tangential skims are often used to determine the area under a smaller peak, located on the leading or tailing edge of a larger peak. With this approach, it is assumed that most of the slope of the baseline is due to the larger peak, and that the concentration of the smaller peak can be represented accurately by the area above the larger peak, as shown in Figure 7.18. This approach tends to underestimate the area of the smaller peak, unless it is very much smaller and narrower than the larger peak.

There are many other options in most software packages to help optimize placement of the baseline. Baseline changes caused by drift or baseline shifts can be compensated for by forcing new baselines to be used. Detection of negative peaks may be inhibited to avoid confusion in changing baselines. Alternatively, routines to detect negative peaks can be activated if appropriate. Valley-to-valley baselines can be used if the peaks are eluted on slopes as a result of gradient elution, or alternatively peaks may be forced by time. The peak detection threshold may be changed for different peaks within a chromatographic run, or the liftoff and touchdown may be changed to compensate for differences in peak widths and heights throughout the chromatogram. The choice of which integration routines to use is a function

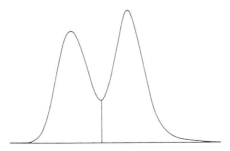

Figure 7.17 Integration of merged peaks by means of a perpendicular drop.

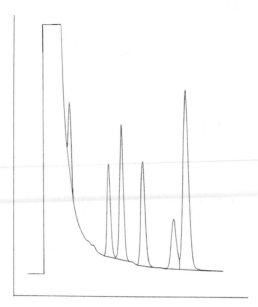

Figure 7.18 Integration of incompletely resolved peaks by means of a tangential skim.

of the chromatogram under consideration and will differ from separation to separation.

7.3.2 Calculation

Integration of a peak is simply the first step in data manipulation for the determination of component concentrations in a sample. Peak integration is performed in order to convert the detector signal into numerical data. There are four principal techniques for determining relative composition information about the sample, all of which rely on the construction of calibration curves. These methods are normalization, the internal standard method, the external standard method, and the method of standard additions.

A calibration curve is a model used to predict the value of an independent variable, the analyte concentration, when only the dependent variable, the analytical response, is known. The normal procedure used to establish a calibration curve is based on a linear least-squares fit of the best straight line for a linear regression, as indicated in

$$Y = mX + b \qquad (7.3)$$

where m is the slope of the curve and b is the value of the intercept of the slope with the Y axis. "Regression" is defined as a functional relationship between two or more correlated variables, and "linear" indicates that the relationship between the two variables is linear. Most analytical chemists prefer linear calibration curves, because the data are easier to interpret than in nonlinear curves.

(i) Normalization

In the normalization method, it is assumed that all the components of the sample have been eluted and detected. However, even when that assumption is invalid, the normalization method always leads to totals representing 100%. In this technique, the area of every peak is measured and summed. The area of an individual peak, X, is then expressed as a percentage of the total peak area, according to

$$\%X = A_X/(A_X + A_Y + A_Z) \times 100 \qquad (7.4)$$

where A_X, A_Y, and A_Z represent the absorbances of the analytes X, Y, and Z, respectively. The normalization approach can be useful when analyzing complex mixtures. However, because the detector response is not the same for all components in the sample, relative peak areas do not necessarily represent the true sample composition. Therefore, a response factor for each component is necessary for normalization. The response factors are calculated relative to a reference compound using

$$F_X = F_r A_r W_X / A_X W_r \qquad (7.5)$$

where A_X and A_r are the areas of the solute and reference peaks, respectively, W_X and W_r are the concentrations of the solute and the reference peaks, respectively, and F_r is the response factor assigned to the reference compound. The corrected area is then obtained by multiplying the area of the peak, A_X, by the relative response factor for that peak, F_X. Although response factors can be calculated from a single measurement, it is advisable to use a calibration method. With the calibration method, a minimum of three different solute concentrations is plotted against detector response and the best straight line drawn among them. The slope of the plot is the response factor. The corrected area is then obtained as above.

(ii) External Standard Method

The external standard method may be applied in a number of ways. The methods operate on the same principle but vary in the amount of work required and therefore in the accuracy of the results. In all cases, standard solutions containing the solutes of interest are prepared, preferably at three different, known concentrations. The standard solutions are then analyzed, and calibration plots of peak area (height) versus concentration are constructed for each solute, as shown in Figure 7.19. From the calibration curves, the unknown concentrations of solutes in the sample can be determined.

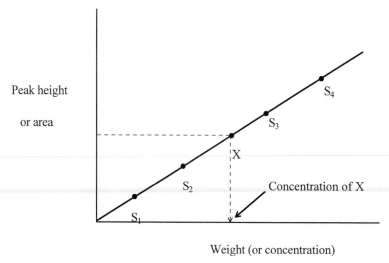

Figure 7.19 Calibration curve for use in external standardization.

In the simplest case, the standards are run once prior to running the samples, in order to construct the calibration curve. All the samples are then run consecutively, and the concentration of the solutes of interest are calculated. This approach is typically used in research and development laboratories, where the identities of the solutes in the samples change regularly. Although more than adequate, this method is the least accurate of the external standard methods.

In most pharmaceutical laboratories concerned with quality control of raw materials such as drugs, the standards are run more often and are used to "bracket" the samples. For example, the operating procedure may involve running three standards followed by six samples followed by the three standards again. A different set of standards may then be used for the analysis of a different formulation, following the same procedure. In this example, the standards on both sides of the samples are used to construct the calibration plots. The intent is to reduce errors resulting from minor changes in the operating conditions, such as air temperature or flow rate fluctuations, that might affect the chromatographic separation.

Another possible scenario is the quality control of a single formulation, day after day. Here, the operating procedure may involve the analysis of two standards followed by three samples, followed by the two standards and three more samples, over and over. A "sliding bracket" would then be used to construct calibration plots. The first two standards and standards 3 and 4 would be used to construct a calibration plot for the first three samples. Standards 3 and 4 would then be used with standards 5 and 6 to construct calibration plots for the next three samples, and so on. The intent is to further reduce errors resulting from minor changes in the operating conditions that might affect the chromatographic separation; this approach provides the most accurate results.

(iii) Internal Standard Method

Internal standards are used to correct for errors in sample preparation or sample introduction and to help determine solute recoveries. They are added to the sample at the earliest possible introduction point in the analytical process. Standards solutions containing the solutes of interest are prepared, preferably at two or three different, known concentrations but with a constant concentration of internal standard. The same concentration of internal standard is also added to each sample. All the standards and samples receive the same treatment from sample preparation, through the sample introduction and separation processes, to detection.

The choice of an internal standard for use in an analytical procedure is important and should be made carefully to avoid the introduction of more errors. The requirements for an internal standard are as follows. (1) It should not be a normal constituent of the sample. (2) It should be completely resolved from all the solutes in the sample. (3) It should be eluted near the solute of interest. (4) It should be added at a concentration that will give a similar peak area (height) as the solute of interest. (5) The physical and chemical properties of the internal standard should resemble those of the solute of interest. (6) It should not react with any of the sample constituents. (7) It should behave in a similar fashion to the other solutes of interest during sample preparation. (8) Finally, it should be of high purity.

Calibration curves are constructed using the standard solutions, as shown in Figure 7.20, by plotting the ratio of the detector response (peak area or peak height) for each solute relative to that of the internal standard against the solute concentration. The peak area ratio for each component in the sample is then calculated, and from these ratios the amount of each constituent in the sample may be determined.

(iv) Standard Addition Method

The method of standard addition is the least widely used method of quantitation, but it is used to ensure that the calibration standards experience the same matrix effects as the sample constituents. In this method, the sample is analyzed first in order to estimate the concentration of the solute(s) of interest. Several different, known concentrations of the solute(s) of interest are then added to portions of the sample, to provide approximate incremental increases in detector response. Each portion of the sample is then reanalyzed. The principle of the method is that the extra signal produced by the addition of the standards is proportional to the original signal. The method is applicable only when a known straight-line relationship has been established between the instrument response and the analyte concentration.[18]

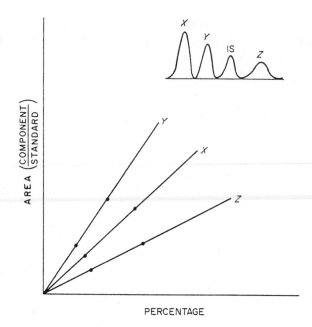

Figure 7.20 Calibration curve for use with internal standardization.

To determine the concentration of the original sample, the detector response is plotted against the concentration of the standard additions, as illustrated in Figure 7.21. As can be seen from Figure 7.21, signal is present when no standard is added; it represents the original concentration of the sample, which is to be determined. To find the original concentration, the straight line is extrapolated to the abscissa (X axis); the absolute value on the abscissa is the original concentration. In actual practice, the use of standard additions is not so simple, since the introduction of additional standards will increase the volume of the sample. Although this approach addresses the sample matrix problem, its main disadvantage is that the concentration of the sample constituents is determined by extrapolation, rather than by interpolation, making this method less precise than other calibration methods. A thorough account has been provided by Bader.[19]

7.3.3 Statistical Treatment of Data

In all measurements there are errors that may be determinate or indeterminate. Indeterminate errors are random errors that cannot be eliminated and are inherent in the analytical technique. When indeterminate errors are minimized, high precision is possible. Determinate errors are errors whose cause and magnitude can be determined. If the determinate errors are minimized, high accuracy can be achieved. Determinate errors include (1) poor sampling techniques, (2) decomposition of the column, (3) change in detector response, (4) poor recorder performance, (5) calculation errors, and (6) operator prejudice or error.

With the increasing use of electronic means of data reduction and the use of computers to generate calibration curves, much of the thought process has been removed from the statistical treatment of chromatographic data. A series of detailed articles has been published discussing concepts in calibration theory,[20-23] and warning against the assumption that a straight line with a correlation coefficient, r, of 1.000 is linear.[24]

The final possible source of determinate errors lies in the plotting and interpretation of the calibration curve. Visual inspection is often the first method used to determine linearity. In addition, most software packages that perform the linear least-squares calculation also calculate the value of the correlation coefficient. In calibration, r ranges from 0 to +1, where 0 indicates no correlation and +1 indicates a perfect positive correlation. The correlation coefficient is not a measure of the quantitative change in one variable with respect to another, but rather a measure of the intensity of association between the two variables.[20] As such, the correlation coefficient may give a high degree of relation between the variables, but the model may give an inadequate fit. Thus, a correlation coeffecient of +1 should not necessarily be interpreted as indicating a perfectly linear correlation.

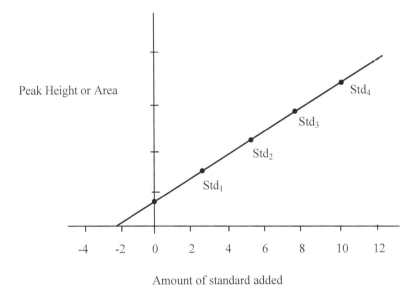

Figure 7.21 Calibration curve for use with standard additions.

As detection limits decrease and calibration curves are generated for wider concentration ranges, care must be taken during evaluation of the curves for linearity. When calibration curves are generated over a wide concentration range, it is common for most of the data points to reside at the lower end of the calibration curve, as illustrated in Figure 7.22. Thus, although the curve may appear to be linear, it is difficult to visually determine linearity at the lower end of the concentration scale.

As a supplemental approach to determine the linearity of calibration curves that span a wide concentration range, many analytical chemists replot the data on a log–log scale. Although that approach makes the data points at the lower end of the concentration scale easier to see, the data are displayed in a nonlinear system which distorts the patterns that existed in a linear scale.[24] The tendency, however, is to interpret the curve using a linear scale, so log–log plots are easily misinterpreted.

There are a number of approaches available to test the linearity of a calibration curve, but probably the most effective is the linearity plot. A linearity plot is a simple procedure for determining the linear working range of a calibration curve,[25] which is obtained by plotting the "sensitivity" at each calibration plot against the standard concentration, as illustrated in Figure 7.23. The sensitivity is calculated by dividing the analytical response at each point on the curve by the concentration. The best horizontal straight line is then drawn through the points (dashed line in Fig. 7.23). The two parallel dotted lines in Figure 7.23 above and below the best-fit line represent the upper and lower acceptable error limits for the analysis, and they are placed on the graph to illustrate the acceptable linear working range. If all the points lie within the limits, the entire data set is considered

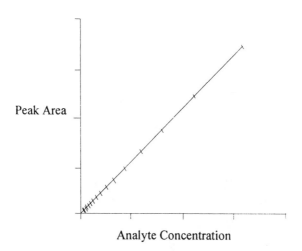

Figure 7.22 Calibration curve covering a wide concentration range.

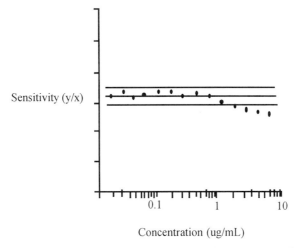

Figure 7.23 Linearity plot for determining the linear working range of a calibration curve.

linear within the acceptable limits. In Figure 7.23, the curve would be considered linear over about two orders of magnitude. Even with this approach, however, potential errors remain. If the calibration curve for the data has a significant Y-intercept, data points that should obviously be outside the limits of the linearity curve may fall within acceptable limits. This error may be addressed by subtracting the value of the Y-intercept from each of the data points prior to calculating the sensitivity.[25]

The temptation to allow data systems to determine the linearity of calibration curves on the basis of the value of the correlation coefficient alone has the potential of introducing determinate errors into chromatographic quantitation. Those errors can be avoided by always studying the results carefully before drawing any conclusions.

7.4 Summary of Major Concepts

1. Data manipulation can be divided into qualitative analysis and quantitative analysis. Qualitative analysis is performed to determine the nature of the analytes in the sample. Quantitative analysis is performed to determine the amount of each analyte in the sample.

2. There are a variety of techniques available to aid in the identification of sample components: matching retention times, standard addition, internal standard, isotopic labeling, enzyme peak shift, and UV and mass spectral libraries.

3. The peak area or height is converted into numerical data by either manual or electronic means. The measurement of peak height is simpler than the measurement of peak area and is often preferred for manual data

reduction. However, peak heights are more susceptible to variations in separation conditions, and the results are not so accurate.

4. There are three measures for manually determining peak areas: product of peak height and peak width at half-height, Condal–Bosch area, and Foley measurement.

5. Geometrical errors associated with manual measurements include height error, peak width error, half-height error, and baseline position error.

6. Three kinds of noise are recognized by the ASTM: short-term noise, long-term noise, and drift.

7. According to ACS guidelines, the limit of detection (LOD) is measured at a signal-to-noise ratio of 3, and the limit of quantitation (LOQ) is measured at a signal-to-noise ratio of 10.

8. Following integration, the analytical response is converted into analyte concentration with the aid of a calibration curve. There are four principal techniques for determining relative composition information about a sample: normalization, internal standard, external standard, and standard addition.

9. Indeterminate errors are errors that cannot be eliminated and are inherent in the analytical technique. Determinate errors are errors whose cause and magnitude can be determined, and they include poor sampling technique, decomposition of the column, change in detector response, improper recorder performance, calculation errors, and operator prejudice or error.

References

1. Zhao, R. H., Li, B. Y., Chen, N., Zhang, Y. K., Wang, Z. Y., and Lu, P. C., *Biomed. Chromatogr.* **7**, 139 (1993).
2. Meyler, M., Kelly, M. T., and Smyth, M. R., *Chromatographia* **36**, 27 (1993).
3. Sewell, P. A., and Clarke, B., *in* "Chromatographic Separations," (D. Kealey, ed.), Wiley, Chichester, 1987.
4. Miller, J. T., "Chromatography: Concepts and Contrasts." Wiley, New York, 1988.
5. Brown, P. R., *J. Chromatogr.* **52**, 257 (1970).
6. Jang, N.-I., and Brown, P. R., *LC-GC.* **10**, 526 (1993).
7. Ryan, T. W., *J. Liq. Chromatogr.* **16**, 1545 (1993).
8. Sims, A., *Amer. Lab.* **June**, 20 (1993).
9. Charley, H., "Food Science." Wiley, New York, 1985.
10. Braithwaite, A., and Smith, F. J., *in* "Chromatographic Methods." 4th Ed., Chapter 8. Chapman & Hall, New York, 1990.
11. Dyson, N., "Chromatographic Integration Methods." Royal Society of Chemistry, Cambridge, 1990.
12. Hamilton, R. J., and Sewell, P. A., "Introduction to High Performance Liquid Chromatography," 2nd Ed. Chapman & Hall, New York, 1982.
13. Majors, R. E., *J. Chromatogr. Sci.* **8**, 338 (1970).

14. Delaney, M. F., *Analyst* **107**, 606 (1982).
15. Ball, D. L., Harris, W. E., and Hapgood, H. W., *Sep. Sci.* **2**, 81 (1967).
16. Ball, D. L., Harris, W. E., and Hapgood, H. W., *Anal Chem.* **40**, 129 (1968).
17. "ASTM E 685-79," 1st Ed. American Society for Testing Materials, Philadelphia, 1981.
18. Bruno, T. J., "Chromatographic and Electrophoretic Methods." Prentice-Hall, Englewood Cliffs, New Jersey, 1991.
19. Bader, M., *J. Chem. Educ.* **57**, 703 (1980).
20. Bonate, P. L., *LC-GC* **10**, 310 (1992).
21. Bonate, P. L., *LC-GC* **10**, 378 (1992).
22. Bonate, P. L., *LC-GC* **10**, 448 (1992).
23. Bonate, P. L., *LC-GC* **10**, 531 (1992).
24. Cassidy, R., and Janoski, M., *LC-GC* **10**, 692 (1992).
25. "Annual Book of ASTM Standards," American Society for Testing Materials, 14.01, pp. 149–158. Philadelphia, 1988.

CHAPTER 8

Miniaturization

8.1 Introduction

Because of high resolution, selectivity, sensitivity, and speed, HPLC and CE have become methods of choice for a vast array of analytical separations. However, with growing environmental concerns over waste disposal, and the high cost of many biological samples, interest in miniaturization is increasing. Miniaturization of the techniques offers many advantages. One advantage is that HPLC, which uses high pressure and high flow rates, can be more readily interfaced to spectroscopic identification techniques such as mass spectrometry, which require much lower flow rates. The interface is enhanced by reducing the size of the LC column, and therefore the rate of flow at which liquid enters the detector. In addition, reducing the size of the column means that the size of the sample can be reduced. Small-bore columns are more economical to manufacture and operate than larger columns, and the reduced consumption of solvent and stationary phase is both economically and environmentally attractive. The most important advantage of the miniaturization of HPLC, however, is the increased mass sensitivity possible with small-bore columns and the reduced sample dilution that occurs on-column.

Interest in the miniaturization of CE has progressed in the direction of CE on a silicon chip (micromachining). This chapter presents an overview of the advantages and disadvantages of various types of miniaturization and the requirements they impose on instrument design. More detailed accounts are provided in the text by Ishii,[1] and in the numerous papers listed at the end of this chapter.

8.2 Classification of Columns

Small-bore columns for HPLC are categorized in terms of the column internal diameter. There are four types of small-bore columns: narrow-bore columns, microbore columns, packed capillaries, and open tubular capillaries. As can be seen in Figure 8.1,[2] the physical features of narrow-bore columns, microbore columns, and the larger inner diameter packed capillaries are similar, except for the column dimensions (Fig. 8.1c). The stationary phase is tightly packed in the column, and the columns are typically slurry packed. Many of these columns are commercially available.

The features of open tubular capillary columns (Fig. 8.1a) and narrow inner diameter packed capillaries (Fig. 8.1b) are significantly different from those of the wider columns. The narrow packed capillaries are much more

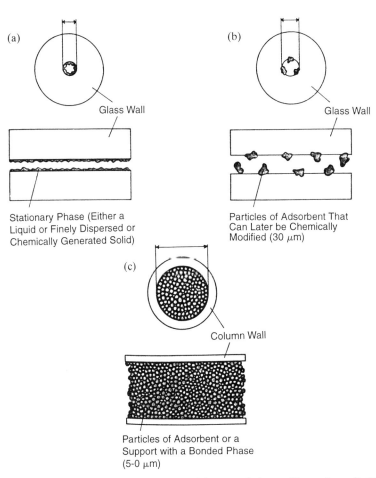

Figure 8.1 Different types of microcolumns: (a) open tubular capillary column (1–50 μm I.D.), (b) dry packed capillary column (≤200 μm I.D.), and (c) microbore column (0.5–1 mm I.D.), narrow-bore column (1–2 mm I.D.), and slurry packed capillary column (≤500 μm I.D.). (Adapted from Ref. 2 with permission.)

loosely packed than are the wider columns. Narrow inner diameter packed capillaries tend to be made by dry packing a standard-bore capillary with the required stationary phase and then extruding the capillary until the required internal diameter is achieved. Open tubular capillaries are prepared by creating a retentive layer on the inside surface of the capillary. Several methods for establishing a retentive layer have been reported in the literature, including chemical modification of etched surfaces,[3,4] immobilization of cross-linked silicones[5–8] and acrylates,[9–11] deposition of liquids,[4] and the realization of a porous silica layer that can be chemically modified.[3,4,12] According to the theory for open tubular liquid chromatography (OTLC), relatively thick stationary layers are required in order to combine the high efficiency and separation speed of narrow columns with sufficient retention and sample capacity. Immobilization of cross-linked silicones and acrylates is most promising for the manufacture of the relatively thick retentive layers required for OTLC.[10]

The preparation of polymer-coated capillaries consists of four consecutive, main steps, as illustrated in Figure 8.2: etching of the bare silica capillary, silylation of the etched surface, *in situ* polymerization, and evaporation of the solvent. Of the four steps, photopolymerization and evaporation of the solvent appear to be the most critical for obtaining uniform layers. The major drawback of the polymeric phases is the poor column efficiency that arises from the small diffusions of solutes in these retentive layers.

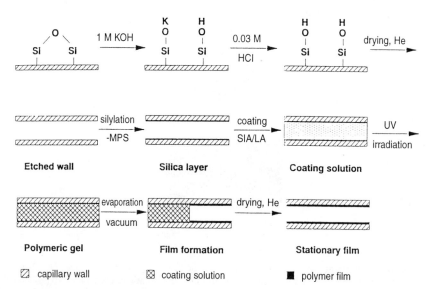

Figure 8.2 Schematic representation of open tubular capillary column preparation. (Adapted from Ref. 10 with permission.)

Table 8.1 Classification of Columns on Basis of Column Inner Diameter[a]

Classification	Column diameter	Reduced diameter (δ)[b]
Open tubular	3–50 μm	0
Packed capillary	≤500 μm	1–100
Microbore column	0.5–1 mm	100–300
Narrow-bore column	1–2 mm	300–600

[a] Adapted from Ref. 13 with permission.
[b] Reduced diameter, δ, is the ratio of column to particle diameters and was introduced by Knox and Parcher.[14]

The range of bore sizes that falls within any given column category is not clearly defined. Several authors have attempted to classify bore sizes definitively.[1,2,13] However, the column diameter recommendations of Sagliano et al.[13] are generally followed, and they are listed in Table 8.1.[14]

8.3 Theoretical and Practical Considerations

The speed of a chromatographic separation is fixed by the particle size, the stationary phase characteristics, the available pressure, the solvent viscosity, the solute diffusivity, the α values of the critical pair, and extra-column dispersion. One way to achieve faster separations is to reduce the particle size of the stationary phase. However, if material of smaller diameter is packed into a conventional size column, the backpressure will become prohibitively high. Thus, in a compromise between speed and optimum performance, narrow (<2 mm) columns packed with small 3–5 μm diameter particles have been developed.

Despite the attraction of high-speed separations, the choice of whether to use standard-bore HPLC or small-bore HPLC is mostly governed by practical considerations (the need for a detector capable of providing positive identifications), environmental considerations (low solvent consumption), sample size considerations (small injection volumes), or price considerations (the need to use exotic solvents and expensive samples, such as DNA). Simple separations are best carried out in conventional short, wide columns, whereas more difficult separations can be achieved with the use of longer, narrower columns.[15] However, difficulties arise when downscaling HPLC. The difficulties are divided into theoretical and practical considerations.

8.3.1 Theoretical Considerations

(i) Sample Volume

Sample volume is the most fundamental factor to be taken into account when designing an HPLC system in which a small-volume column is used, because it determines most of the instrumental requirements.[1] The maximum sample volume, V_s, that can be injected into a chromatography system, without introducing band broadening or a loss in column efficiency, can be expressed according to

$$V_s = \frac{d^2 L \varepsilon (1 + k')\pi C}{4(N^{1/2})} \quad (8.1)$$

where d is the column diameter, L is the column length, ε is the porosity of the column packing material, k' is the capacity factor, C is a constant, and N is the number of theoretical plates. Thus, V_s is directly proportional to the length of the column and the square of the column diameter. The constant, C, which is dependent on the injection technique, is approximately 4 under most experimental conditions.[16] See Ref. 1 for more detailed information on sample volume.

The concept of maximum sample volume as it relates to miniaturization can be illustrated as follows, assuming that both a conventional analytical column (4.6 mm I.D.) and a narrow-bore column (2 mm I.D.) have equal efficiency, length, and porosity. To switch a method from the analytical column to the narrow-bore column and still maintain optimum performance on the narrow-bore column, Eq. (8.1) may be rewritten as

$$\frac{(V_s)_{\text{micro}}}{(V_s)_{\text{conv}}} = \frac{d^2_{\text{micro}}}{d^2_{\text{conv}}}. \quad (8.2)$$

Thus, V_s should be 4.8 times smaller to ensure a similar column performance and to avoid any loss in column efficiency. Similarly, the sample volume should be 21 times smaller for a microbore column with an inner diameter of 1 mm.

Reducing the sample volume, however, means that the mass of sample loaded onto the narrow-bore column is reduced, and therefore the mass of sample reaching the detector is decreased. This imposes difficulties on the detection scheme because less sample is available for detection. Therefore, it is important that the entire instrument be optimized for use with small-bore columns. Examples of reduced sample volumes for each of the column categories are listed in Table 8.2.

(ii) Dilution Factor

Although chromatography is a separation process, dilution occurs during the process of separation. The small sample volume that is injected onto the head of the column disperses in the mobile phase during passage through the column. The dilution factor can be expressed in terms of the maximum peak concentration, C_{max}, and is related to the injected sample mass, m_s, according to

$$C_{\text{max}} = \frac{2(2^{1/2}) m_s (N^{1/2})}{d^2 L \varepsilon (1 + k') \pi^{3/2}} \quad (8.3)$$

8.3 Theoretical and Practical Considerations

Table 8.2 Comparison of Sample Volumes for Various Classes of Columns, Relative to Conventional Standard-Bore Column (4.6 mm I.D.)

Classification	Column diameter	Sample volume reduction
Narrow bore	2.1 mm	4.8×
Microbore	1 mm	21×
Packed capillary	320 μm	207×
Open tubular	20 μm	32,900×

where, as before, d is the column diameter, L is the column length, ε is the porosity of the column packing material, and k' is the capacity factor. The maximum peak concentration is inversely proportional to the length of the column and the square of the column diameter. Therefore, by decreasing the column diameter, the maximum peak concentration may be increased.

This concept can be illustrated in the following manner. If the same sample mass is injected onto a conventional column and a narrow-bore column of equal length, porosity, and efficiency, Eq. (8.3) may be written as

$$\frac{(C_{max})_{micro}}{(C_{max})_{conv}} = \frac{d^2_{conv}}{d^2_{micro}} \qquad (8.4)$$

Reducing the column diameter from 4.6 to 2 mm, results in a theoretical fivefold increase in peak concentration. This means that the injected sample is approximately 20% less diluted and an increase in detectability can be expected. The increased peak concentration obtainable by decreasing the column inner diameter, with all other parameters equal, is balanced by the decreased sample volume that should theoretically be injected. The only way to see the increased peak concentration, under these conditions, is to concentrate a large sample volume on the head of the column prior to running the separation.

(iii) Flow Rate

To downscale accurately from an existing and well-established chromatography method using standard-bore columns and obtain a chromatogram with a similar profile, it is important to take into account the flow rate. When downscaling, the flow rate must be adjusted in order to maintain a constant linear velocity. The relationship between linear velocity, u, and flow rate, F, is expressed according to

$$u = \frac{4F}{\pi d^2 \varepsilon}. \qquad (8.5)$$

Thus, in order to maintain the same linear velocity for columns of different diameters, the flow rate should be decreased in proportion to the square of the ratio of the diameters. This concept can be illustrated as follows, for a method using a conventional column (4.6 mm I.D.) operating at 2.5 ml/min that is to be downscaled to a narrow-bore column (2.1 mm I.D.). The ratio, r, of the square of the diameters is

$$r = \frac{(4.6)^2}{(2.1)^2}$$

$$= 4.8.$$

Substituting this ratio into Eq. (8.5), all other parameters being equal, gives

$$4.8 = \frac{(2.5 \text{ ml/min})}{F_2}$$

$$F_2 \cong 0.5 \text{ ml/min}.$$

Thus, the same chromatographic separation can be obtained with standard-bore and narrow-bore columns, but the flow rate must be adjusted to maintain a constant linear velocity to achieve similar retention times and peak response.

8.3.2 Instrumental Considerations

Chromatography is a dilution process. In addition to the dispersion that occurs during passage of the solute through the column, extracolumn effects also contribute to peak broadening. Extracolumn dispersion is the contribution to peak variance that takes place outside the column. If it is not minimized, extracolumn dispersion may significantly degenerate the separation that has been previously obtained in the column. When using small-bore columns, where peak volumes are less than in standard-bore systems, it is even more important to minimize extracolumn contributions.

There are four major sources of extracolumn dispersion: (i) dispersion due to the injection volume, (ii) dispersion due to the volume of the detector cell, (iii) dispersion due to the detector response time, and (iv) dispersion resulting from the volume in the connecting tubing between the injector and the column and also between the column and the detector. Thus, extracolumn dispersion takes place between the injector and the detector, only, and the system volume contributed by the solvent delivery system does not contribute to dispersion. The total permitted extracolumn dispersion (variance) is shared, albeit unequally, between those dispersion sources. A commonly accepted criterion for the instrumental contribution to zone broadening, suggested by Klinkenberg,[17] is that it should not exceed 10% of the column variance.

Commercially available HPLC instrumentation was originally designed for use with standard-bore columns (4.6 mm I.D.). Detector flow cells were optimized for maximum sensitivity with these analytical columns, injectors were designed to introduce microliter quantities of sample, and pumps were designed to be accurate and reproducible in the milliliter flow-rate ranges commonly employed with standard-bore columns. However, these instruments are not well suited for use with small-bore columns, as the dispersion introduced by the large volumes is detrimental to the separation. In addition, the reproducibility and accuracy of the pumping system at the low flow rates required are questionable.

Because of the interest in narrow-bore and microbore columns, instrument manufacturers now have developed solvent delivery systems that are capable of accurately pumping at the low flow rates typically required for microbore applications (>10 μL/min). In addition, injectors have been designed that are capable of introducing the smaller sample volumes, and detector cells are available that are small enough to monitor the reduced sample volumes passing through the detector. Thus narrow-bore and microbore applications are possible with readily available instrumentation, and reports may be found in the literature.

For capillary LC, most of the literature is still being produced by academic research groups, as instrumentation sufficiently miniaturized in terms of detector flow cells and solvent delivery systems is not available. The solvent delivery system, which includes not only the pump but also components such as check valves, pulse dampers, pressure transducers, and mixing chambers, is one of the most important components of the liquid chromatograph, and performance of the solvent delivery system directly affects the retention time reproducibility and detector baseline stability. Stable, reproducible solvent delivery systems capable of operating in the nanoliter per minute flow-rate range can be made by interfacing a conventional microscale pump with a flow-splitting device that is commercially available.

(i) Injection Volume

Ideally, a sample is introduced into a chromatograph as a perfect plug. In practice, this is not the case, and diffusion occurs because of the injector. For narrow-bore and microbore applications, injectors capable of introducing the required sample volumes are commercially available and optimized to reduce dispersion. This is not the case for capillary LC, and homemade injection systems include the sample tube technique, in-column injection, stopped-flow injection, pressure pulse-driven stopped-flow injection (PSI), groove injection, split injection, heart-cut injection, and the moving injection technique (MIT). Of the injection techniques, only the split injector, MIT and PSI approaches can introduce subnanoliter sample volumes accu-

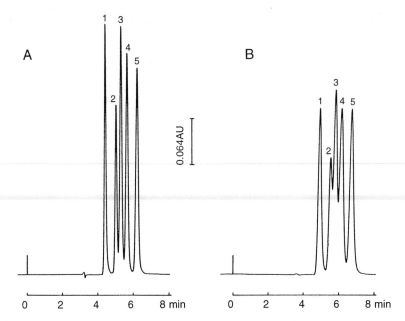

Figure 8.3 Chromatograms obtained with different injectors. (A) Chromatogram obtained using a microvalve injector; (B) chromatogram obtained using a conventional injector. Conditions: column, C_{18} (250 mm × 1.5 mm I.D.); mobile phase, acetonitrile/water (90/10 v/v); flow rate, 0.1 ml/min; injection volume, 1 ml; detection, UV absorbance at 250 nm with 1-ml cell; sample solvent: acetonitrile. Peaks: 1, benzene; 2, naphthalene; 3, biphenyl; 4, fluorene; 5, anthracene. (Reprinted from Ref. 1 with permission.)

rately. For large capillaries with inner diameters exceeding 40 μm, and even MIT is not very reproducible. A comparison of the simple, cost-effective split injection device with the more sophisticated and expensive PSI system in terms of band broadening and injection volume reproducibility for 5-μm open tube capillaries, indicated that both systems were capable of providing high efficiency but that with the PSI system larger volumes were injected more accurately. The effect of the injector on the separation can be seen clearly in Figure 8.3.

If the sample is dissolved in a solvent that is weaker than the mobile phase, then the sample can be enriched on the head of the column without penetrating into the column bed. This compression effect is particularly important for capillary LC applications, since it permits significantly larger injection volumes. A substantial increase in sensitivity results, and conventional autosamplers with 20-μl loops can be used.[16] However, sample solubility and recovery, miscibility of the sample with the mobile phase, and the maximum tolerable loss in column efficiency and resolution must all be assessed experimentally for optimum on-column focusing.[16]

8.3 Theoretical and Practical Considerations

(ii) Detector Cell Volume

To detect a narrow band from a small-volume column, the detector cell must be small enough to prevent diffusion. The detector variance increases as the flow rate increases. Practically speaking, as long as the detector cell volume is less than about one-tenth of the volume of the peak of interest, extracolumn broadening by the detector will be insignificant.[3] If those dimensions are used, even if complete remixing of the solute band occurs in the detector cell, the increase in peak volume is less than 8%.[1] Thus, for a 4.6-mm-I.D. standard-bore column, with a peak volume of 100 μl, the maximum cell volume is 10 μl. For a 0.5-mm-I.D. microbore column, with a peak volume of 1 μl, the maximum cell volume is 0.1 μl. Obviously, as the column dimensions decrease, the maximum cell volume for optimum performance becomes prohibitively small. The effect of using an excessively large detector cell volume is illustrated in Figure 8.4.

To maximize the peak response, it is important to use the longest cell path length possible without introducing extracolumn dispersion. With

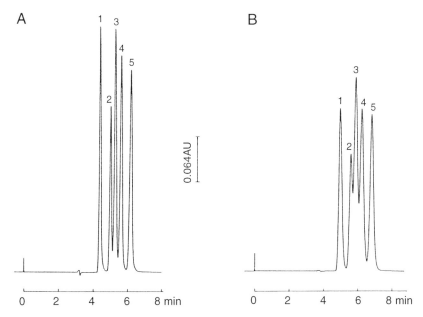

Figure 8.4 Effect of detector cell volume on separation efficiency: (A) 1-μl cell; (B) 8-μl cell. Conditions: column; 250 mm × 1.5 mm I.D. C_{18} (5 μm); mobile phase: acetonitrile (90%)/water (10%); flow rate, 100 μl/min; injection volume, 1 μl; detection, UV absorbance at 250 nm. Peaks: 1, benzene; 2, naphthalene; 3, biphenyl; 4, fluorene; 5, anthracene. (Reprinted from Ref. 1 with permission.)

spectrophotometric detection, the maximum signal, A, generated by the spectrophotometer is given by the Lambert–Beer law:

$$A = \varepsilon c l \qquad (8.6)$$

where l is the cell path length, ε is the molar absorptivity of the solute, and c is the solute concentration. From the Beer–Lambert equation [Eq. (8.6)] it can be seen that the signal is directly proportional to the detector cell path length. Decreasing the volume of the detector cell by reducing the length of the detector cell, therefore, will result in a loss in sensitivity. If the path length is kept constant and the cell volume is reduced by decreasing the width of the cell, however, the optical transmittance of the cell will decrease because of the decreasing cell aperture width, resulting in poor linearity and higher noise levels. Nevertheless, within limits, decreasing the detector cell width tends to provide better sensitivity than does decreasing the detector cell length. In either case, smaller detector cell dimensions result in higher limits of detection. Figure 8.5 shows the effect on sensitivity of increasing the volume of the detector cell; both the peak response and the dispersion increase. Figure 8.5A shows the best resolution but the worst response, whereas Figure 8.5C shows the best response with the worst resolution; Figure 8.5B shows the best compromise.

(iii) Detector Time Constant

The detector time constant is the response time of the detector to the signal passing through it. A slower time constant will result in less apparent noise, but it will compromise signal and also resolution for closely eluting peaks. For closely eluting peaks, therefore, it is important to use a faster time constant. If there is plenty of resolution, a slower time constant will provide a smoother baseline. The effect of the time constant is illustrated in Figure 8.6.

(iv) Connecting Tubing

Both the dimensions of the tubing and the manner in which it is connected are of utmost importance when using small columns with small injection volumes, small detector cell volumes, and low flow rates. The larger the amount of tubing, in terms of both diameter and length, the greater is the opportunity for dispersion. In addition, any diffusion effects will be magnified in the small volumes. Tubing that is not seated properly at the bottom of the connection holes, or tubing that has been cut at an angle, so that it cannot be seated flush with the bottom of the holes, will have a significant effect on the separation efficiency. It is important, therefore, when using microbore or capillary columns, to use the narrowest tubing possible and to ensure that it is cut so that the end of the tubing is perfectly perpendicular to its length. The effect of improperly seating the tubing is illustrated in Figure 8.7.

8.3 Theoretical and Practical Considerations

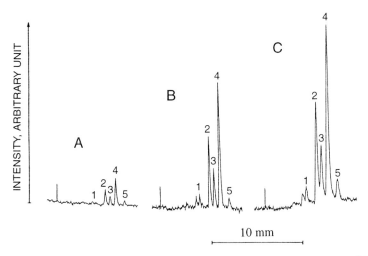

Figure 8.5 Chromatograms for same sample amount using detector cells with different volumes: (A) 1.5 μl; (B) 3.0 μl; (C) 7.0 μl. Conditions: column, 250 mm × 1.5 mm I.D. C_{18} (10 μm); injection volume, 1 μl; detection, spectrophotometric (λ_{ex} 252 nm, λ_{em} 360 nm). Peaks: 1, benzene; 2, naphthalene; 3, biphenyl; 4, fluorene; 5, anthracene. (Reprinted from Ref. 1 with permission.)

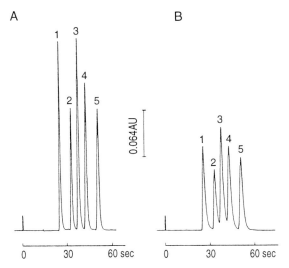

Figure 8.6 Influence of time constant on peak shape and sensitivity: (A) 0.2 sec; (B) 2 sec. Column, 50 mm × 4.6 mm I.D. C_{18} (3 μm). Peaks: 1, α-tocopherol; 2, β-tocopherol; 3, γ-tocopherol; 4, δ-tocopherol. (Adapted from Ref. 1 with permission.)

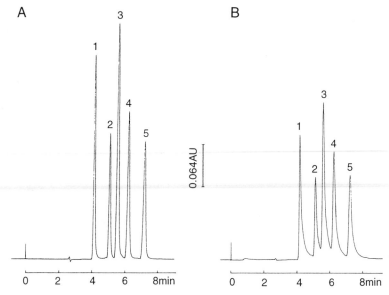

Figure 8.7 Effects of poor connecting tubing on separation efficiency: (A) properly connected tubing; (B) tubing seated improperly at either end of the column. Conditions: column, 250 mm × 1.5 mm I.D. C_{18} (5 μm); mobile phase, acetonitrile (90%)/water (10%); injection volume, 1 μL; detection, UV absorbance at 250 nm; detector cell volume, 1 μL. Peaks: 1, benzene; 2, naphthalene; 3, biphenyl; 4, fluorene; 5, anthracene. (Reprinted from Ref. 1 with permission.)

8.4 Applications of Miniaturized Liquid Chromatography

Despite the advantages of downscaling HPLC, neither narrow-bore nor microbore LC has become as popular as might have been expected. Instead, capillary LC using both packed capillaries and open tubular capillaries is the area of most intense research. Nevertheless, microbore and narrow-bore applications continue to appear in the literature, especially in situations where limited sample volumes restrict the scale of the separation.

A review of HPLC methods for antiepileptic drug analysis was published in 1987 by Juergens.[18] Standard analytical separations were compared with narrow-bore separations, and a 70% reduction in the cost of solvents was possible, owing to a reduction in flow rate from 1 ml/min for the analytical column to 0.3 ml/min for the narrow-bore column. Gayden et al.[19] developed an isocratic method, using narrow-bore columns, to quantify adenosine (Ado) release by dispersed rat renal outer medullary cells under conditions of normoxia and hypoxia. Standard HPLC with UV detection has been the predominant method for studying the metabolic pathways of adenosine and measuring the Ado breakdown products inosine (Ino) and hypoxanthine (Hyp). However, the conventional methods lack reliability

8.4 Applications of Miniaturized Liquid Chromatography   255

when quantifyng Ado, Ino, and Hyp at a concentration range of 3–5 pmol, the typical concentrations found in isolated cells or cell suspensions. The instrumentation the investigators used consisted of a commercially available pump with microflow modifications, a standard injector with a 10-μl loop, and a UV detector equipped with a 2.4-μl flow cell. The method was linear over the range tested (3–50 pmol), with minimum limits of quantitation being 2.9 ± 0.2 pmol for Ado, 4.2 ± 0.3 pmol for Ino, and 4.9 ± 0.4 pmol for Hyp. Figure 8.8 shows chromatograms for the three compounds prepared in the mobile phase and also in the cell suspension medium.

The 1989 European Community Water Act restricts the concentration of all pesticides in drinking water (<500 ng/liter), and therefore sensitive,

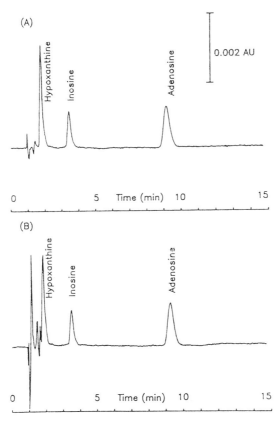

Figure 8.8 Retention times for adenosine (22.45 pmol), inosine (22.37 pmol), and hypoxanthine (44.08 pmol): (A) standard prepared in mobile phase; (B) standard prepared in incubation fluid or cell suspension medium. Conditions: mobile phase, 125 mM potassium dihydrogen phosphate, 1.5% (v/v) acetonitrile, 20 mM triethylamine, and 1.0 mM tetrabutylammonium hydrogen sulfate (TBAHS) (pH 6.5); flow rate, 0.5 ml/min. (Reprinted from Ref. 18 with permission.)

reproducible methods for the analysis of pesticides in water are required. An application of microbore HPLC for the determination of phenylurea herbicides in water with UV and electrochemical detection was developed to address the problem.[20] Microbore HPLC was chosen instead of standard analytical HPLC because of the advantages it offered in terms detection limits and solvent consumption. The column was a reversed-phase C_{18} column (30 cm × 1.0 mm), and the mobile phase consisted of 10 mM lithium perchlorate in 75:25 (v/v) methanol/water (adjusted to pH 5.5 using 1% phosphoric acid). A comparison of the sensitivity of UV detection and electrochemical detection for a group of pesticides is shown in Figure 8.9. Microbore HPLC has been applied to many areas of analysis, including therapeutic drug monitoring and toxicology[21,22] and analysis of pharmaceutical compounds,[23] aromatic hydrocarbons,[24,25] physiological fluids,[24] and numerous nitrogen-containing compounds.[25]

Capillary HPLC is an area of considerable interest. Applications for packed capillary LC include analysis of human hormones,[26] aromatic hydrocarbons,[27] short-chain hydrocarbons,[28] polynuclear aromatic hydrocarbons,[29] pesticides,[30] physiological fluids,[31] mapping and sequencing of enzymatic digests,[32] and cephalosporin antibiotics.[33] Chervet et al.[31] modified commercially available instrumentation using a microflow processor in order to obtain a flow rate of 6 μl/min. Comprising a conventional autosampler equipped with a 20-μl sample loop, a variable-wavelength UV detector with a 20-mm, 90-nl Z-shaped capillary flow cell, and a 15 cm × 320 μm I.D. column packed with 3-μm C_{18} packing material, the system was applied to the analysis of plasma extracts for benzodiazepines. A large sample volume was used, but the sample was concentrated onto the head of the column, thereby minimizing extracolumn dispersion arising from the injector. In comparing results to those with a conventional 4.6-mm-I.D. column, the authors noted an increase in sensitivity of 25–30 times, which was increased to 50- to 60-fold when the capillary diameter was reduced to 180 μm I.D. In addition a 99% decrease in solvent consumption was obtained with the 180-μm-I.D. capillary compared with the 4.6-mm-I.D. column. Figure 8.10 shows a comparison of the sensitivity obtained with a conventional 4.6-mm-I.D. column versus the 320-μm-I.D. capillary.

Interfacing liquid chromatography with mass spectrometry is becoming more practical with the miniaturization of HPLC equipment, and more reports are appearing in the literature. Even so, the reports are still mostly from academic institutions where homemade interfaces are developed. Cappiello and Bruner[26] developed an interface to couple the effluent from an open tubular liquid chromatograph (OTLC) to the commercially available desolvation chamber and momentum separator of a particle beam mass spectrometer. Micro flow rates were obtained using a flow-splitting device, and the sample was injected using an injector equipped with a 60-nl internal loop. Capillary columns were made using 250-μm-I.D. Polyethylethylke-

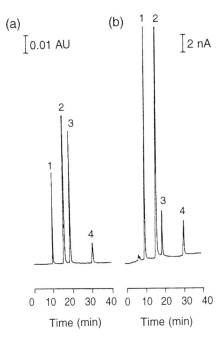

Figure 8.9. Separation of four phenylurea herbicides using (a) on-line UV detection at 254 nm and (b) on-line electrochemical detection with 1.35 V oxidation potential. Peaks: 1, metoxuran; 2, diuron; 3, linuron; 4, neburon. (Reprinted from Ref. 20 with permission.)

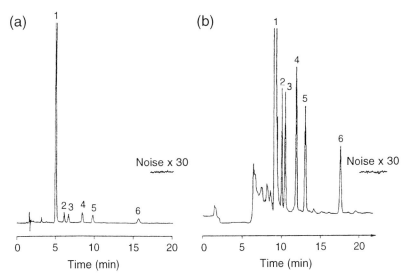

Figure 8.10 Separation of benzodiazepines in plasma extracts using (a) conventional HPLC (4.6 mm I.D.) and (b) capillary LC (320 μm I.D.). Conditions: mobile phase, 35:10:55 (v/v/v) acetonitrile/methanol/10 mM phosphate buffer (pH 7.0); flow rate, 6.0 μl/min; injection volume, 20 μl; detection, UV absorbance at 313 nm. Peaks: 1, carbamazepine; 2, nitrazepam; 3, clonazepam; 4, flunitrazepam (internal standard); 5, nordazepam; 6, diazepam. (Reprinted from Ref. 31 with permission.)

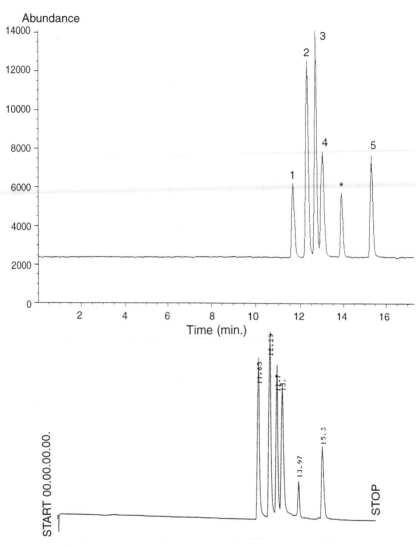

Figure 8.11 Separation of a mixture of human hormones: (a) total ion current (TIC) profile; (b) UV profile (254 nm). Conditions: injection volume, 60 nl; capillary, 25 cm × 250 μm I.D. PEEK capillary, C_{18}, 5 μm; flow rate, 1 μl/min; gradient, 40–80% acetonitrile in 25 min. Peaks: 1, nortestosterone; 2, testosterone; 3, diethylstilbestrol; 4, methyltestosterone; 5, medroxyprogesterone; *, diethylstilbestrol impurity. (Reprinted from Ref. 26 with permission.)

tone (PEEK) tubing packed with C_{18} reversed-phase material (5-μm particle size). The instrument was tested using a sample of human hormones. Figure 8.11 shows the total ion current (TIC) profile and the UV (254 nm) profile of five hormones. No appreciable difference in peak shape or width is apparent between the two chromatograms, indicating that no extra dispersion was caused by the modified mass spectrometry interface.

Open tubular LC has been applied to the analysis of compounds such as polynuclear aromatic hydrocarbons,[10,11,34] inorganic anions,[35] inorganic cations,[35] amino acids,[36] antitumor drugs,[36] catechols,[37,38] anilines,[39] aromatic amines,[40] phthalates,[40] and nucleosides.[40] An open tubular liquid chromatograph has even been designed on a silicon chip, although its utility is untested.[41] Detection is typically on-column and includes fluorescence, UV absorbance spectroscopy, mass spectrometry, and electrochemical detection. Of the four classes of small-bore columns, open tube capillaries offer the most advantages in terms of reduced solvent consumption, compatibility with mass spectrometry, efficiency, and mass sensitivity. However, because of the instrumental restrictions imposed by the small sample volumes, the OTLC technique is the most difficult to perform using commercially available instrumentation. Swart et al.[11] prepared and evaluated a series of polyacrylate-coated fused silica capillaries for reversed-phase OTLC. Figure 8.12 shows chromatograms of anthracene derivatives obtained using different modes of detection. The high mass loading capacity of the thick polyacrylate coatings allows on-column UV detection to be performed, substantially increasing the applicability of OTLC compared with the more specialized and sensitive detectors, such as laser-induced fluorescence and mass spectrometry.

8.5 Separations on Chips

The microfabrication of analytical instrumentation provides an interesting alternative for chemical sensing. With sample handling, separation, and detection methods all incorporated into a single, small probe, the device could resemble a sensor in many regards, although each function would be under the control of the user. The combination of all sample handling and measurement steps into a single, standard-sized package, incorporating a high level of automation, is known as total chemical analysis.[42–44] The systems constructed for continuous monitoring of samples are called total chemical analysis systems (TAS), or μ-TAS in a miniaturized format.[45]

Although the manufacture of a liquid chromatographic chip containing a capillary column and a detector has been reported,[41,46,47] most research has been directed toward the development of CE on a chip,[48–55] as the experimental design of the CE technique is simple and resolution is dependent on the electric field strength rather than on the length of the capillary. With small capillary lengths and high voltages, high-speed separations are possible. In addition, the devices are small and readily transportable.

8.5.1 Fabrication of Planar Devices

Microfabrication technologies using photolithographic patterning processes (micromachining) were used first by Terry[56,57] in 1975 for the integra-

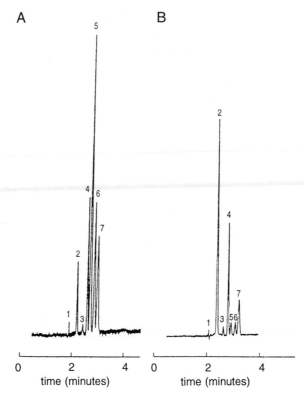

Figure 8.12 Chromatography of anthracene derivatives with different detection modes: (A) laser-induced fluorescence; (B) UV absorbance. Conditions: capillary, 115.1 cm × 10 μm I.D. with 1.27-μm coating; stationary phase, silicone acrylate/ethylhexyl acrylate; $V_S/V_M = 0.65$; mobile phase, acetonitrile; detection, (A) laser-induced fluorescence (λ_{ex} 325 nm, λ_{em} 380 nm), pressure 13.8 bar, (B) UV (258 nm), pressure 13.0 bar. Peaks: 1, salicylate; 2, anthracenemethanol; 3, anthracenecarbonitrile; 4, anthracene; 5, fluoranthracene; 6, 1,2-benzanthracene; 7, 9-phenylanthracene. (Reprinted from Ref. 11 with permission.)

tion of a gas chromatograph onto a silicon wafer. It was not until 1992,[48,49] however, that micromachining was applied to the integration of an electrophoresis capillary and sample injector on a planar device. The steps involved in micromachining a device for CE, which are shown diagramatically in Figure 8.13, involve film deposition (Fig. 8.13a), photolithography (Fig. 8.13b,c), etching (Fig. 8.13d–f), and bonding (Fig. 8.13g).

As illustrated in Figure 8.13 the CE capillary channel is etched into the bottom plate and a second plate is bonded on top. The channel dimensions are on the order of 10–12 mm deep and 30–70 mm wide. Most designs are rectangular in shape, although trapezoidal channels have also been constructed that are 70 μm wide at the bottom of the channel and 90 μm wide at the top. Although those dimensions are the most typical, larger channels ranging up to 1 mm have been created.

In some chip designs, the top plate has electrodes defined in it, and holes are drilled at various places in order to allow access to the etched channels. Pipette tips are glued into the holes to serve as reservoirs. In another design,[55] constructed on a microscope slide, the channel is etched onto a microscope plate and then covered with the circular coverslip. Following the bonding, cylindrical plastic reservoirs are affixed to the chip using epoxy; thus, no hole drilling is necessary. The design is illustrated in Figure 8.14. The glass plates are thermally bonded at around 600–650°C, but it has been suggested by Jacobson et al.[55] that direct bonding would be more practical for higher melting point substrates, such as quartz. Direct bonding involves hydrolyzing two glass plates in dilute NH_4OH/H_2O_2, prior to annealing them at 500°C.

The sample is introduced into the separation channel by electromigration injection. Thus far, detection has been solely by laser-induced fluorescence, using an argon ion laser or a helium/neon laser as the excitation source and a photomultiplier tube (PMT) or charge-coupled device (CCD)

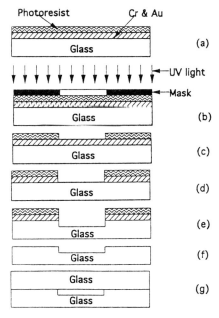

Figure 8.13 Sequence of photolithographic fabrication: (a) Cr- and Au-masked glass plate coated with photoresist, (b) plate exposed to light through a master mask, (c) photoresist developed, (d) exposed metal mask etched, (e) exposed glass etched, (f) resist and metal stripped, and (g) glass cover plate bonded to form capillary. (Reprinted from Ref. 54 with permission.)

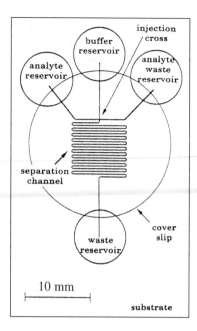

Figure 8.14 Planar device in which reservoirs are affixed to the chip following bonding of the two plates, and no holes are drilled into the top plate. The large circle represents the coverslip, and the smaller circles represent the reservoirs. (Reprinted from Ref. 55 with permission.)

camera to detect the emitted radiation. Reproducibility has been reported to be on the order of 3% or better for peak area[50,53] and better than 1% for migration times.[50,53] Channels containing bonded polyamide,[58] polyacrylamide,[59] and C_{18}[60] coatings have been reported, and analytes have been separated by CZE,[53] MECC,[61,62] gel electrophoresis,[62] and CEC.[60]

8.5.2 Sample Introduction

A typical design of a planar device for CE is shown in Figure 8.15. The sample is placed in reservoirs 1 and 2, and buffer is placed in reservoirs 3 and 4 (Fig 8.15). The sample is introduced into the system by electromigration injection, but as long as the injection time is sufficiently long, the sample bias typical of electromigration injection in conventional CE is not significant. When a voltage is applied between reservoirs 1 (positive) and 2 (ground), the sample migrates along the sample inlet channel (Fig. 8.15). Initially, the concentration of faster migrating ions passing through the intersection of the two channels is higher than the concentration of the slower migrating species. As the voltage continues to be applied, however, the slower migrating species also pass through the intersection, and the composition of the sample at the intersection is therefore fairly representa-

8.5 Separations on Chips

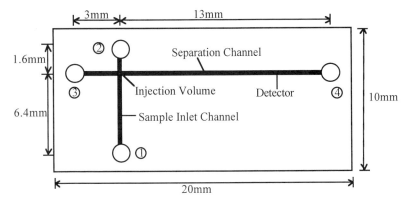

Figure 8.15 Simple planar device for CE. (Redrawn from Ref. 54 with permission.)

tive of the original sample. As long as a voltage is applied between reservoirs 1 and 2, a finite, defined volume of sample will be present at the intersection.

When a voltage is applied between reservoirs 3 (positive) and 4 (ground), the buffer migrates along the separation channel (Fig 8.15). Sample present at the intersection is swept into the separation channel by the movement of the buffer, and separation then occurs in the separation channel. The device shown in Figure 8.15 has been used to separate amino acids labeled with fluorescein isothiocyanate (FITC), namely, labeled arginine (Arg-FITC), phenylalanine (Phe-FITC), and glutamic acid (Glu-FITC), within 3 sec using a potential of 2500 V between reservoirs 3 and 4.

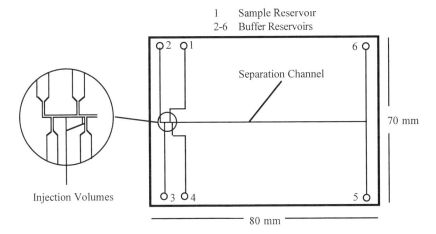

Figure 8.16 Planar CE device with different injection volumes. (Redrawn from Ref. 53 with permission.)

Using the device shown in Figure 8.15, the sample volume is fixed. One way to introduce different sample volumes is by etching more channels on the glass plate. This concept is illustrated in Figure 8.16. Reservoir 1 is filled with sample, while the remaining reservoirs are filled with buffer. When a voltage is applied between reservoirs 1 (+2 kV) and 4 (ground), sample travels from reservoir 1 to reservoir 4 (Fig. 8.16), thereby filling a

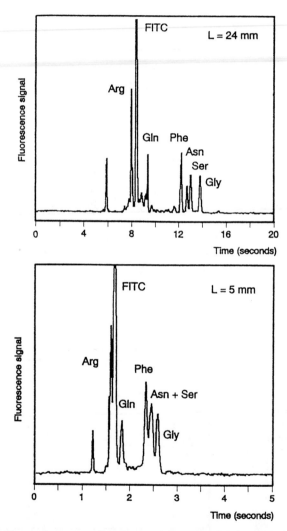

Figure 8.17 Electropherograms of a mixture of six FITC-labeled amino acids recorded at separation lengths of (a) $L = 24$ mm and (b) $L = 5$ mm. The electric field strength in both cases was 1060 V/cm, and the formal concentration of each amino acid was 10 mM. The buffer solution was 20 mM boric acid/100 mM Tris (pH 9.0). (Reprinted from Ref. 53 with permission.)

8.5 Separations on Chips

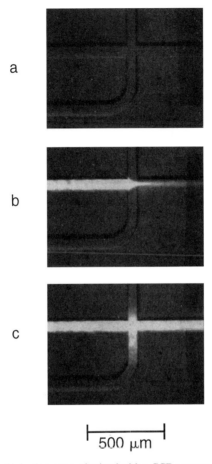

Figure 8.18 Images of injection cross, obtained with a CCD camera, (a) with no fluorescent analyte present, (b) during pinched sample loading procedure, and (c) during floating sample loading procedure. (Reprinted from Ref. 55 with permission.)

150-μm section of the separation capillary. To alter the volume of sample introduced, voltage is applied between reservoir 1 and a reservoir other than reservoir 4. For example, when a voltage is applied between reservoirs 1 and 3, the sample travels from reservoir 1 to reservoir 3, thereby filling a 300-μm section of the separation capillary. To effect a separation, voltage is applied between reservoirs 2 and 5, thereby sweeping the sample already present at the beginning of the separation channel along toward the detector. The device illustrated in Figure 8.16 has been used to separate six FITC-labeled amino acids. The detector was positioned at different places along the separation channel, thereby altering the effective length of the

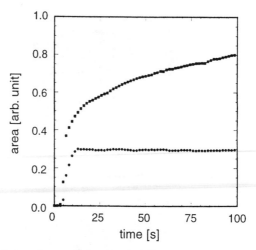

Figure 8.19 Variation of amount of analyte in injection area with time for pinched sample loading (●) and floating sample loading (■) using rhodamine B. (Reprinted from Ref. 55 with permission.)

channel. When the effective length, L, was 24 mm, all six amino acids were baseline resolved. At $L = 5$ mm, Asn-FITC and Ser-FITC comigrated, but the rest of the amino acids were mostly resolved. The separation is shown in Figure 8.17.

The devices illustrated in Figures 8.15 and 8.16 have had linear separation channels etched into the glass, and thus the length of the separation channel has been severely limited by the dimensions of the chip. Separation channel geometries are not limited to straight channels, however, and a serpentine geometry has also been shown to be effective,[55] as illustrated in Figure 8.14. Jacobson et al.[55] created a planar CE device in which a 165-mm-long separation channel was contained in an area of less than 10 × 10 mm. It was concluded that band broadening phenomena associated with the serpentine structure, while measurable, did not appear to be a severely limiting problem for CE implementations employing field strengths less than about 200 V/cm.

Jacobson et al.[55] investigated the use of a different format of electromigration injection in an attempt to increase the reproducibility of the injection procedure. Using the device shown in Figure 8.14, they compared the standard injection procedure described above, which they called floating sample loading, with a second procedure, which they termed pinched sample loading. The effect of the two procedures is illustrated in Figure 8.18. As can be seen from Figure 8.18c, with the standard floating injection procedure, sample diffuses into the separation channel during the loading stage, resulting in irreproducible injection volumes.

To perform the floating injection procedure using the device shown in Figure 8.18, a voltage is applied between the analyte reservoir and the analyte waste reservoir (ground). To perform the pinched injection procedure, potentials are also applied to the buffer reservoir and waste reservoir, with the analyte waste reservoir grounded. To perform the separation, a potential is applied to the buffer reservoir, with the waste reservoir grounded and with the analyte and analyte waste reservoirs at approximately half the potential of the buffer reservoir. The difference in reproducibility of the two injection modes is shown in Figure 8.19. Thus, by using the pinched injection mode, the contribution of the injection mode to diffusion can be minimized.

8.5.3. Synchronized Cyclic Capillary Electrophoresis

Multiple column switching has been made possible by the construction of a cyclic CE microstructure in which four channels that are 20 mm in length are arranged to form a square, as illustrated in Figure 8.20.[62] Two channels at each corner connect the loop to external reservoirs filled with separation buffer. Platinum electrodes are inserted into the reservoirs for the application of high voltages. One side of the device has a volume-defined injection scheme incorporated into it. The overall technique is called synchronized cyclic capillary electrophoresis.[63]

The purpose of synchronized cyclic CE is to provide a high degree of

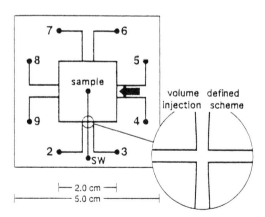

Figure 8.20 Layout of cyclic CE chip and volume-defined injection scheme. The squared separation channel has a circumference of 80 mm, widths of 40 and 20 μm at the top and bottom, respectively, and a depth of 10 μm. The volume of the injection scheme is approximately 12 pl. SW refers to sample waste, and the arrow marks the point of LIF detection. Reservoirs 3, 5, 6, and 8 and SW are reservoirs into which channel effluent is pumped electrokinetically. Reservoirs 2, 4, 7, and 9 and "sample" are reservoirs from which electrolytes are introduced into the channel system. (Reprinted from Ref. 62 with permission.)

flexibility with respect to the total capillary length, thereby customizing the apparatus for any given sample mixture. In addition, cyclic instead of single-pass microchip instrumentation permits the application of relatively low voltages and the repetitive detection of single peaks. The principle of synchronized cyclic CE is illustrated in Figure 8.21,[62] optimized for the detection for sample component 2. Following sample introduction (Fig. 8.21A), a separation voltage is applied between reservoirs 2 and 6 (see Fig. 8.20). The sample components travel toward a waste reservoir and begin to separate (Fig. 8.21B). When component 2 reaches the corner, component 3 has already entered the waste channel (Fig. 8.21C). At this point the voltage is switched to reservoirs 4 and 8 where it is held until sample component 2 reaches the next corner (Fig. 8.21D). The voltage is switched to reservoirs 7 and 3 during which time sample component 1 leaves the loop (Fig. 8.21E), and finally the voltage is applied between reservoirs 9 and 5 and sample component 2 is redetected before flowing to waste (Fig. 8.21F).

An MECC immunoassay for serum theophylline was achieved on the cyclic CE microstructure, one to two orders of magnitude faster (~30 sec) than the comparable separation using conventional fused silica capillaries, with higher efficiency and at no expense to accuracy or precision. The glass channels were not coated for these experiments, but they were rinsed with

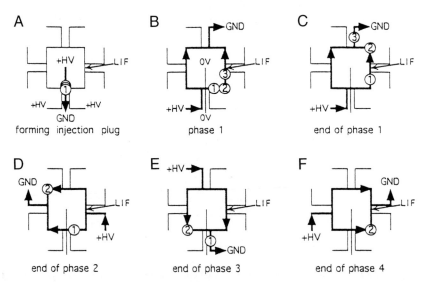

Figure 8.21 Principle of synchronized cyclic CE. Separation of three sample components (1, 2, and 3) over one cycle is illustrated. The direction of fluid flow and sample transport is indicated by the arrows; LIF refers to the location of the detector. The voltage switching protocol is synchronized to cycle component 2: (A) injection phase, (B) time point during phase 1, (C) end of phase 1, (D) end of phase 2, (E) end of phase 3, and (F) end of phase 4 and end of first cycle. (Reprinted from Ref. 62 with permission.)

0.1 M sodium hydroxide (1 min) and buffer (1 min) between runs. The chip was also shown to give comparable separations of FITC-labeled amino acids by MECC and gel electrophoresis.

The T junctions and corners of the layout currently used in synchronized cyclic CE contribute more to zone broadening than do the channel connections at the corners used in single-pass chips. In addition, a significant percentage of analyte is lost on each cycle as it passes by T junctions and the detector. These issues are under investigation.

8.5.4 Separation Efficiency

A major focus of researchers creating planar CE devices is speed of analysis, in part so that the devices can be used as chemical sensors and circumvent the severe selectivity and lifetime requirements of conventional chemical sensors. To increase the speed of analysis, shorter capillaries should be used, in combination with higher electric field strengths. Optimum efficiency depends on minimization of all unavoidable sources of band broadening, in addition to the elimination of nonideal effects such as Joule heating and adsorption on capillary walls. Therefore, work to understand the contributions which limit the efficiency of the separation is continuing.

The most important contributions to band broadening in CE, and ultimately the factors limiting the separation efficiency, are longitudinal diffusion, on-column and off-column contributions arising from the injected and detected volumes, and the detector response time. Several authors[64-66] have investigated the influence of sample injection on resolution and quantitation in standard CE, pointing out the crucial role it plays for the performance of CE. If the efficiency is expressed in terms of the plate height, H_{tot}, then the primary contributions to band broadening are given by

$$H_{tot} = H_{inj} + H_{det} + H_{diff} + H_{geo} \qquad (8.7)$$

where H_{inj}, H_{det}, H_{diff}, and H_{geo} correspond to the contributions to the plate height from the injection plug length, detector path length, molecular diffusion in the axial direction, and geometry of the channel, respectively. For a linear channel H_{geo} is negligible, but it is measurable for a serpentine geometry; for field strengths below about 200 V/cm, however, it does not appear to be a severely limiting problem.[55]

The contribution of axial diffusion to the band broadening follows the Einstein equation:[67]

$$H_{diff} = 2D_m/\mu \qquad (8.8)$$

where D_m and μ are the diffusion coefficient of the analyte and the linear velocity of the analyte, respectively. Axial diffusion can be minimized by performing the analysis in the minimum time possible.

The contributions of the injection plug and detection path length can be calculated by[68]

$$H = l_{inj}^2/(12L) \quad (8.9)$$

$$H = l_{det}^2/(12L) \quad (8.10)$$

where l_{inj} and l_{det} are the channel lengths of the injection plug and detection path, respectively, and L is the length of the separation channel between the point of injection and the point of detection. Thus, by injecting the sample using the pinched injection mode and by minimizing the length of the injection plug and of the detection path, while increasing the length of the separation channel, the efficiency can be maximized.

8.6 Summary of Major Concepts

1. Small-bore columns for HPLC are categorized in terms of the column internal diameter. There are four types of small-bore columns:

narrow-bore columns,
microbore columns,
packed capillaries, and
open tubular capillaries.

2. The column inner diameters for the four classes of small bore columns are

narrow bore, 1–2 mm;
microbore, 0.5–1 mm;
packed capillary, ≤500 μm; and
open tubular, 3–50 μm.

3. When transferring a method from a standard-bore column (4.6 mm I.D.) to a small-bore column, the sample volume injected is decreased according to the equation

$$\frac{(V_s)_{micro}}{(V_s)_{conv}} = \frac{d_{micro}^2}{d_{conv}^2}$$

where the subscript micro refers to the smaller column and conv refers to the conventional column.

4. To downscale accurately from an existing method using standard-bore columns and still obtain a similar profile, the flow rate should be adjusted according to Eq. (8.5), in order to maintain a constant linear velocity:

$$u = \frac{4F}{\pi d^2 \varepsilon}$$

where F is the flow rate, u is the linear velocity, d is the column diameter, and ε is the porosity of the column packing.

5. It is particularly important to minimize extracolumn dispersion when using small-bore columns because the peak volumes are sufficiently small that extracolumn dispersion can degenerate the separation significantly. There are four major sources of extracolumn dispersion:

dispersion due to the injection volume,
dispersion due to the volume of the detector cell,
dispersion due to the detector response time, and
dispersion resulting from the volume in the connecting tubing between the injector and the column and also between the column and the detector.

6. Systems constructed for continuous monitoring of samples are termed total chemical analysis systems (TAS), or μ-TAS when in a miniaturized format. Both HPLC and CE have been manufactured on a chip. Research has been directed mainly toward the development of CE on a chip, however, because the experimental design of the CE technique is simple and resolution is dependent on the electric field strength rather than on the length of the capillary.

7. The ability to perform column switching on a CE chip is called synchronized cyclic CE.

References

1. Ishii, D., ed., "Introduction to Microscale High-Performance Liquid Chromatography." VCH Publ. New York, 1988.
2. Novotny, M., *Anal. Chem.* **53**, 1295A (1981).
3. Tock, P. P. H., Stegeman, G., Peerboom, R., Poppe, H., Kraak, J. C., and Unger, K. K., *Chromatographia* **24**, 617 (1987).
4. Tock, P. P. H., Boshoven, C., Poppe, H., Kraak, J. C., and Unger, K. K., *J. Chromatogr.* **477**, 95 (1989).
5. Folestad, S., and Larsson, M., *J. Chromatogr.* **394**, 455 (1987).
6. Gohlin, K., and Larsson, M., *J. Microcolumn Sep.* **3**, 547 (1991).
7. van Berkel, O., Kraak, J. C., and Poppe, H., *J. Chromatogr.* **499**, 345 (1990).
8. Dluzneski, P. R., and Jorgenson, J. W., *HRC & CC, J. High Resolut. Chromatogr. Chromatogr. Commun.* **11**, 332 (1988).
9. Eguchi, S., Tock, P. P. H., Kloosterboer, J. G., Zegers, C. P. G., Schoenmakers, P. J., Kraak, J. C., and Poppe, H., *J. Chromatogr.* **516**, 301 (1990).
10. Ruan, Y., Feenstra, G., Kraak, J. C., and Poppe, H., *Chromatographia* **35**, 597 (1993).
11. Swart, R., Kraak, J. C., and Poppe, H., *J. Chromatogr.* **670**, 25 (1994).
12. Scott, R. P. W., *J. Chromatogr. Sci.* **23**, 233 (1985).
13. Sagliano, N., Jr., Hsu, S.-H., Floyd, T. R., Raglione, T. V., and Hartwick, R. A., *J. Chromatogr. Sci.* **23**, 238 (1985).
14. Knox, J. H., and Parcher, J. F., *Anal. Chem.* **41**, 1599 (1969).

15. Scott, R. P. W., *J. Chromatogr. Sci.* **23**, 233 (1985).
16. Chervet, J. P., van Soest, R. E. J., and Salzmann, J. P., *LC-GC* **10**, 866 (1992).
17. Klinkenberg, A., *in* "Gas Chromatography 1960" (R. P. W. Scott, ed.), Butterworth, London, 1960.
18. Juergens, U., *J. Liq. Chromatogr.* **10**, 507 (1987).
19. Gayden, R. H., Watts, B. A. III., Beach, R. E., and Benedict, C. R., *J. Chromatogr.* **536**, 265 (1991).
20. Boussenadji, R., Dufek, P., and Porthault, M., *LC-GC* **11**, 450 (1993).
21. Wong, S. H. Y., Marzouk, N., Aziz, O., and Sheeran, S., *J. Liq. Chromatogr.* **10**, 491 (1987).
22. Wong, S. H. Y., Cudny, C., Aziz, O., Marzouk, N., and Sheeran, S., *J. Liq. Chromatogr.* **11**, 1143 (1988).
23. Tsuji, K., and Binns, R. B., *J. Chromatogr.* **253**, 227 (1982).
24. Scott, R. P. W., *J. Chromatogr. Sci.* **18**, 49 (1980).
25. Taylor, L. T., *J. Chromatogr. Sci.* **23**, 265 (1985).
26. Cappiello, A., and Bruner, F., *Anal. Chem.* **65**, 1281 (1993).
27. McGuffin, V. L., and Novotny, M., *J. Chromatogr.* **255**, 381 (1983).
28. Takeuchi, T., Murayama, M., and Ishii, D., *Chromatographia* **25**, 1072 (1988).
29. Ishii, D., and Takeuchi, T., *J. Chromatogr.* **255**, 349 (1983).
30. Lee, E. D., and Henion, J. D., *J. Chromatogr. Sci.* **23**, 253 (1985).
31. Chervet, J.-P., van Soest, R. E. J., and Salzmann, J. P., *LC-GC* **10**, 866 (1993).
32. Kassel, D. B., Shushan, B., Sakuma, T., and Salzmann, J.-P., *Anal. Chem.* **66**, 236 (1994).
33. Moore, C. M., Sato, K., and Katsumata, Y., *J. Chromatogr.* **539**, 215 (1991).
34. Crego, A. L., Diez-Masa, J. C., and Dabrio, M. V., *Anal. Chem.* **65**, 1615 (1993).
35. Muller, S., Scheidegger, D., Haber, C., and Simon, W., *J. High Resolut. Chromatogr.* **14**, 174 (1991).
36. Balchunas, A. T., Capacci, M. J., Sepaniak, M. J., and Maskarinec, M. P., *J. Chromatogr. Sci.* **23**, 381 (1985).
37. Knecht, L. A., Guthrie, E. J., and Jorgenson, J. W., *Anal. Chem.* **56**, 497 (1984).
38. St Claire, R. L. III, and Jorgenson, J. W., *J. Chromatogr. Sci.* **23**, 186 (1985).
39. Kucera, P., and Guiochon, G., *J. Chromatogr.* **283**, 1 (1984).
40. Ishii, D., and Takeuchi, T., *J. Chromatogr. Sci.* **18**, 462 (1980).
41. Manz, A., Miyahara, Y., Miura, J., Wanatabe, Y., Miyagi, H., and Sato, K., *Sens. Actuators* **B1**, 249 (1990).
42. Widmer, H. M., *Trends Anal. Chem.* **2**, 8 (1983).
43. Widmer, H. M., Ererd, J.-F., and Grass, G., *Int. J. Environ. Anal. Chem.* **18**, 1 (1984).
44. Graber, N., Ludi, H., and Widmer, H. M., *Sens. Actuators* **B1**, 239 (1990).
45. Manz, A., Graber, N., and Widmer, H. M., *Sens. Actuators* **B1**, 244 (1990).
46. Cowen, S., and Craston, D. H., *in* "Micro Total Analysis Systems" (A. van den Berg and P. Bergveld, eds.), pp. 295–298. Kluwer Academic Publishers, Dordrecht, The Netherlands, 1995.
47. Ocvirk, G., Verpoorte, E., Manz, A., Grasserbauer, M., and Widmer, H. M., *Anal. Methods Instrum.* **2**, 74 (1995).
48. Harrison, D. J., Manz, A., Fan, Z., Ludi, H., and Widmer, H. M., *Anal. Chem.* **64**, 1926 (1992).

49. Manz, A., Harrison, D. J., Verpoorte, E. M. J., Fettinger, J., Paulus, A., Ludi, H., and Widmer, H. M., *J. Chromatogr.* **593**, 253 (1992).
50. Seiler, K., Harrison, D. J., and Manz, A., *Anal. Chem.* **65**, 1481 (1993).
51. Harrison, D. J., Fluri, K., Seiler, K., Fan, Z., Effenhauser, C. S., and Manz, A., *Science* **261**, 895 (1993).
52. Harrison, D. J., Glavina, P. G., and Manz, A., *Sens. Actuators* **B10**, 107 (1993).
53. Effenhauser, C. S., Manz, A., and Widmer, H. M., *Anal. Chem.* **65**, 1926 (1993).
54. Fan, Z. H., and Harrison, D. J., *Anal. Chem.* **66**, 177 (1994).
55. Jacobson, S. C., Hergenroeder, R., Koutny, L. B., Warmack, R. J., and Ramsey, J. M., *Anal. Chem.* **66**, 1107 (1994).
56. Terry, S. C., A gas chromatography system fabricated on a silicon wafer using integrated circuit technology. Ph.D. Dissertation, Stanford University, Stanford, California, 1975.
57. Terry, S. C., Jerman, J. H., and Angell, J. B., *IEEE Trans. Electron Devices* **ED-26**, 1880 (1979).
58. Pukl, M., Prosek, M., and Kaiser, R. E., *Chromatographia* **38**, 83 (1994).
59. Jacobson, S. C., Moore, A. W., and Ramsey, J. M., *Anal. Chem.* **67**, 2059 (1995).
60. Jacobson, S. C., Hergenroeder, R., Koutny, L. B., and Ramsey, J. M., *Anal. Chem.* **66**, 2369 (1994).
61. Moore, A. W., Jacobson, S. C., and Ramsey, J. M., *Anal. Chem.* **67**, 4184 (1995).
62. von Heeren, F., Verpoorte, E., Manz, A., and Thormann, W., *Anal. Chem.* **68**, 2044 (1996).
63. Burggraf, N., Manz, A., Verpoorte, E., Effenhauser, C. S., Widmer, H. M., and de Rooij, N. F., *Sens. Actuators* **B20**, 103 (1994).
64. Dose, E. V., and Guiochon, G., *Anal. Chem.* **64**, 123 (1992).
65. Delinger, S. L., and Davis, J. M., *Anal. Chem.* **64**, 1947 (1992).
66. Huang, X., Coleman, W. F., and Zare, R. N., *J. Chromatogr.* **480**, 95 (1989).
67. Giddings, J. C., "Dynamics of Chromatography, Part I: Principles and Theory," Chapter 2. Dekker, New York, 1965.
68. Sternberg, J. C., *Adv. Chromatogr.* **2**, 205 (1966).

Index

A
Absorbance detectors
  CE, 194–196
  HPLC, 91–97
Accelerated solvent extraction, 115–116
Adsorption, 2
Adsorption chromatography, see Normal phase chromatography
Affinity chromatography, 3, 50–66
Amperometric detectors
  CE, 200–202
  HPLC, 102, 104–105
Ampholytes, 174–175
Analog-to-digital converter, 227
Apparent mobility, 137–138, 144, 150
Applied potential, 137
Area, measurement of 223–228
Autosamplers, 84–85, 116

B
Band broadening, 8, 12–13
  see also Eddy diffusion Longitudinal diffusion, Rate theory and Mass transfer, resistance to
Baseline drift, 91, 125, 229, 230
Baseline noise, 125–126, 229
Baseline placement, 231–232
Beer's law, 92, 98, 252
Bimodal set, 48
Biospecific elution, 55
Bonded-phase, 26, 27
Bonded phase chromatography see Reversed phase chromatography
Bubble cell, 193

Buffer systems in CE, 155–157, 159–160, 163–164, 171–174, 177, 179–181

C
Calibration curves, 232–239
Capacity, 7, 8, 11, 20
Capacity factor, optimization of, 8–12
Capillaries for CE, 136, 158, 190–193
  conditioning, 207
Capillary electrochromatography, 170–172
Capillary electrophoresis, 134
Capillary gel electrophoresis, 166–170
Capillary isoelectric focusing, 174–178
Capillary isotachophoresis, 178–182
Capillary wall, 142, 149, 192
Capillary zone electrophoresis, 155–161
Carbon load, 30–31
Carrier ampholytes, see Ampholytes
Carrier electrolyte, see buffer
CE in a chip, 260
Charge-to-mass ratio, 154
Chemiluminescence detection, 108
Chiral separations
  CE, 172–176
  HPLC, 57–62
Chiral stationary phases, 58–61
Chromatography, 1
Chromophores, 92, 93
Coelectroosmosis, 136
Columns
  care and use, 89–90
  construction materials, 89

275

Columns (*continued*)
  internal diameter, 88
  efficiency, 31
  length, 88
  particle size, 88
  stationary phase, 85–86
Column selectivity, 31
Column switching, 116–118
Competition model, 25
Compositional accuracy, 81–82
Condal–Bosch area, 224–226
Conductivity detectors
  CE, 199–200
  HPLC, 102–104
Convective transport, 15
Cooling, capillary, 185–186
Correlation coefficient, 237
Counter-electroosmosis, 136
Critical micelle concentration, 36, 161
Current, troubleshooting, 208

D

Degassing, *see* Solvent degassing
Derivatization, 100–102, 172
  capillary wall, 142
Detectors
  cell volume, 250–252
  classification, 90
  CE, 194–205
  HPLC, 91–108
  properties, 90–91, 194
  time constant, 252
  variance, 19
Dialysis, 110, 111
Diffusion, *See* Rate theory and Electromigration dispersion
Diffusive transport, 15
Digitization, 227
Dilution factor, 246–248
Dispersion, 248–254
  see also Electromigration dispersion
Displacement chromatography, 4, 7, 20
Donnan exclusion chromatography, *see* Ion-exclusion chromatography
Dynamic ion-exchange model, 33

E

Eddy diffusion, 15, 16–17, 21, 145
Effective mobility, 135

Efficiency, 7, 11, 13, 19, 20, 85, 143–150
  see also Zone broadening, Rate theory and Plate theory
Electrochemical detectors
  CE, 199–200
  HPLC, 102, 104–105
Electrodialysis, 110
Electroendoosmosis, *see* Electroosmotic flow
Electrolyte systems, 135
Electromigration dispersion, 145–147
Electromigration injection, 187–189
  in planar devices, 263–267
Electroosmotic flow, 138–143, 150, 155, 162–163, 165, 168, 170, 174, 177–178, 189
  control in capillaries, 192
  definition, 138
  factors affecting, 139–141
  manipulation of, 141–143
  measurement of, 139
Electrophoresis, 134
Electrophoretic mobility, 136, 137, 138, 189
  definition, 137
Electrophoretic mobilization, 178
Elution development, 3–7, 20
Enantiomers, *see* Chiral separations
Enzyme peak shift, 217–220
End-capping, 31
EOF, *see* Electroosmotic flow
Exclusion limit, 47
External standard, 233–235
Extra-column band broadening, 19–20

F

Field-amplified sample injection, 189–190
Field programming, 186–187
Filtration, 110, 111, 118
Floating injection, 266–267
Flow accuracy, 81, 83
Flow-rate, 72–73, 77–78, 121, 248
Fluorescence detector
  CE, 196–199
  HPLC, 98–102
Foley's equation, 224, 225–227

Fraction collection, 205
Fractionation range, 47
Frontal analysis, 4, 7, 20
Fronting peak

G

Gaussian peaks, 13
Gel-filtration chromatography, 46
    see also Size exclusion
        chromatography
Gel-permeation chromatography, 46
    see also Size-exclusion
        chromatography
Gradient elution, 5–7
Gradient pumping systems, 80–83
Guard column, 116

H

H or HETP, 13
Height, measurement of, 222, 223
High-performance liquid
    chromatography
    definition, 1
    basic concepts, 7–20
High pressure mixing, 80–81
High voltage power supply, 186–187
Hjerten S., 146
Holdup time, 8
HPLC, see High-performance liquid
    chromatography
Hydrodynamic mobilization, 178
Hydrophobic interaction
    chromatography, 24, 38, 64–66
Hydrostatic injection, 187

I

Indirect detection, 94, 97
Injection length, 145
Injection valve, 83–84
Injection volume, 19, 250
Internal standard, 216–217, 235
Ion chromatography, 40, 43–44
Ion exchange capacity, 43
Ion-exchange chromatography, 3, 38, 43–46, 64–66
Ion-exclusion chromatography, 34
Ion-interaction chromatography, see
    Ion pair chromatography

Ion interaction model, 34
Ionization methods, for mass
    spectrometry, 106, 107
Ion-pair chromatography, 33–35
Ion pair mode, 33
Ion pair reagent, 33, 34
Ion-suppression chromatography, 33
Isocratic elution, 4–5
Isoelectric point, 154, 175
Isotopic labeling, 217

J

Joule heating, 147–148

L

Lambert Beer law, see Beer's law
Leading electrolyte, 178, 179
Liquid chromatography
    definition, 1
    classification, 2–7
Limit of detection, 229–230
Limit of quantitation, 230
Linear dynamic range, 91
Linearity plot, 238–239
Liquid chromatography, classification, 2–7
Liquid–solid chromatography, see
    Normal-phase chromatography
Liquid solvent extraction, 114–116
Longitudinal diffusion
    CE, 145
    HPLC, 15, 17, 21
Low pressure mixing, 81–83
Lukacs K.D., 148

M

Martin A.J.P., 13
Mass spectrometric detector
    Ce, 202–205
    HPLC, 105–108
Mass transfer, resistance to, 15, 17–18, 21
MECC, see Micellar electrokinetic
    capillary chromatography
Mechanism of separation, 25
Membrane devices for sample
    preparation, 109–110, 111
Mesh size, 170

Metal chelation, 173
Metal complexation, 35–36
Micellar chromatography, 36
Micellar electrokinetic capillary
    chromatography, 161–166, 268
Microbore, 72, 243
Micromachining, 261–263
Microwave extraction, 115
Migration, 137
Migration time, 137, 143–144, 150, 209
Minimum detectable quantity, see
    Limit of quantitation and Limit of
    detection
Mobile phase
  HPLC, 10–11, 26–27, 31–32, 38,
    43–44, 48–49, 55, 60, 61–62, 104
Mobility, 147

N

Neutral marker, 139
Noise, see Baseline drift; Baseline
    noise detector properties and
    signal-to-noise ratio
Non-specific elution, 55
Normal-phase chromatography, 24–29
Normalization, 233
Nuclear magnetic resonance detector,
    108
Number of theorectical plates, see
    Theoretical plates and Rate
    theory

O

Ogston model, 166
Ohmic heating, see Joule heating
Ohm's law plot, 148, 149
Open tubular liquid chromatography,
    244

P

Paired ion chromatography, see Ion
    pair chromatography
Partition chromatography, 2–3
Partition ratio, see Capacity factor
Partitioning theory, 30
Peak at half height method, 14

Peak identification, see Qualitative
    analysis
Peak measurement, see Quantitative
    analysis
Peak shape, 124, 145–147
PEEK, for HPLC pumps, 74
Pellicular materials, 87–88
Perfusive transport, 15
Permeation limit, 47
Photolithographic fabrication, 261–262
Planar devices, 261–267
Plate height, see H or HETP
Plate number, see Theoretical plates
Plate theory, 13–15, 21
Polyacrylamide gel, 168–169, 192
Polyethylethylketone, see PEEK
Polymer networks, 166, 168, 169, 192
Polytetrafluoroethylene, see Teflon
Pore size, 87–88
Porous particles, 87
Pressure accuracy, 81, 83
Pulse dampers, 77
Pumps, 72–77, 80–83, 170
    see also Solvent delivery system
Purification factor, 56

Q

Qualitative analysis, 214–222
Quantitative analysis, 222–239

R

Radioactivity detector, 108
Raman spectroscopy, detection,
    108–109
Rate equation
Rate theory, 13, 15–20, 21
Reciprocating-piston pumps, 75–76
Recovery, 56
Refractive index detector, 90, 108
Relative retention, see also Selectivity
    factor
Reptation model, 166
Resistance to mass transfer, see Rate
    theory
Resolution, 7, 11–12, 21, 49, 85, 123,
    150, 158, 163, 178
Response factor, 233

Retention mechanism, 2–3, 25, 29–30, 38, 40–41, 46–47, 50–52, 57–58, 60
Retention time, 8, 214–216
Retention volume, 8
Reversed-phase chromatography, 29–37, 64–66
see also sample preparation
Reverse osmosis, 110, 111
Ripple, 81, 82

S

Sample introduction
  CE, 187–190, 263–267
  HPLC, 83–85
Sample concentration, 189–190
Sample preparation,
  CE, 205–206
  HPLC, 109–118
Sample stacking, 189
Selectivity, 7, 9–11, 20
Selectivity coefficient, 41
Sensitivity, see Detectors
Separation factor, see Selectivity factor
SFC, see Supercritical fluid chromatography
Sieving media, 168–169
5 Sigma method, 14
Signal-to-noise ratio, 91
Silylation, 31
Size-exclusion chromatography, 3, 46–50
  see also Gel filtration chromatography and Gel permeation chromatography
Solvent degassing, 78–79
Solid phase extraction, 110–113, 116–118
Solvent delivery systems, 71–78, 249–250
Solvent extraction, 115–116
Solvent interaction model, 25
Solvent polarity parameter, 32
Solvent selectivity, 26
Solvent solubility parameter, 27
Solvent strength, 27
Solvophobic theory, 29

Soxhlet extraction, 114
Spacer arm, 54
Spiking, see Standard addition
Sparging, 78–79
Stainless steel, for HPLC pumps, 73
Standard addition, 216, 235–236
Stationary phase, 25–26, 30–31, 38, 41–43, 47–48, 52–55, 58–61, 172
Step gradient, 7
Supercritical fluid extraction, 114, 115
Suppressed conductivity detection, 200
Suppressor, 40
Surface enhanced Raman spectroscopy detection, 108–109
Surfactants, 36, 142, 161, 163–164, 166
Synchronized cyclic capillary electrophoresis, 267–269
Synge, see Martin
Syringe pumps, 74–75
System evaluation

T

Tailing peak, 146–147, 224
Tangent method, 14
Tangential skim, 231–232
Teflon, for HPLC pumps, 74
Theoretical plates, 13
  see also H and Plate theory
Time constant, see Detector time constant
Titanium, for HPLC pumps, 73–74
Trailing electrolyte, 178, 179
Tubing, extracolumn effects, 19, 252–254

U

Ultrafiltration, 110, 111
Ultraviolet/visible absorbance detector
  CE, 194–195
  HPLC, 91–97

V

Vacuum degas, 78
Van Deemter, 13, 15
Van Deemter equation, 15
Van Deemter plot, 15–16
Velocity, 135, 137, 138

Void volume, 9
Voltage effects, 141, 144, 148, 150, 158–159

W
Water, effects on normal phase stationary phases, 27

Z
Z-cell, 193
Zeta potential, 139–140
Zone broadening
Zwitterionic compounds, *see* Capillary isoelectric focusing